SP Pvt. Ltd.

Biotechnology Series: **Vol. 1**

ANIMAL CELL CULTURE

Bina Mishra, B.P. Mishra, Pran P. Bhat and P.N. Bhat

Centre for Integrated Animal Husbandry Dairy Development,
Flat No. 205, Block No. F – 64/C9,
Sector – 40, Noida – 201 301

2011

Studium Press (India) Pvt. Ltd.

Biotechnology Series: Vol. 1

ANIMAL CELL CULTURE

Series ISBN: 978-93-80012-49-0
ISBN: 978-93-80012-50-6

Published by:

Studium Press (India) Pvt. Ltd.
4735/22, 2nd Floor, Prakash Deep Building
(Near Delhi Medical Association),
Ansari Road, Darya Ganj, New Delhi-110 002
Tel.: +91-11-43240200-15 (15 lines); Fax: +91-11-43240200
jngovil@gmail.com; jngovil@hotmail.com

Printed at:
Salasar Imaging Systems
Delhi-110035 (India)

ABOUT THE SERIES

It was realized by 1985 that the available diagnostics and vaccines used in Animal Health Protection were not giving the kind of protection which was required for the country, the size of India, it was also realized that in addition to artificial insemination, country needed to undertake research and development in embryo transfer and associated technologies in cattle for which technologies had been developed and standardized in scientifically advanced countries. These needed to be researched in buffaloes, a species contributing 65% of the total milk in the country, and standardized particularly when this could be used as an adjunct to progeny testing programes in livestock species.

In order to enhance the genetic improvement in cattle and buffaloes and other species of livestock the embryo transfer and associated technologies were considered vital and critical if the gains obtained from crossbreeding of cattle were to be sustained.

Dr S. Ramachandran who was then the Chairman, National Biotechnology Board (NBTB), Department of Science and Technology in Ministry of Science and Technology, was convinced that if animal biotechnology had to assist India to sort out the problems of enhancement of food, we had to put a two pronged attack to solve the problems, he used to tell that after all the Americans are not going to solve the problems of basic reproduction in buffaloes. It is our animal and we got to solve these problems. The board considered several projects on the basic and applied animal biotechnology including a mission mode project for Embryo Transfer Technology (ETT) and associated reproductive technologies.

By 1986 the Cabinet approved the creation of a new science department of biotechnology with Dr S. Ramachandran as the secretary. Dr Ramachandran established a task force on animal biotechnology in Department of Biotechnology (DBT), which was charged with responsibility of creating the structural plan for the Animal Biotechnology Sector with Prof. (Dr) P.N. Bhat, as the Chairman. Concurrently it was realized that there was total absence of manpower in the sector and except for a few laboratories, on biotechnology applications no research or development was being done in the country, hence it was necessary that apart from introducing short term training programmes, it was necessary to initiate a master's level programe in biotechnology with specific emphasis on application. One such centre was established at Indian Veterinary Research Institute (IVRI) as a part of a National Centre on Animal Biotechnology along with R & D Project on Embryo transfer technology for training the manpower. All the short term training programes had mandatory arrangements for producing training manuals. A similar project on producing training manuals for master's degree programmes was also initiated.

These manuals have been in use since 1987 in various areas of animal biotechnology at IVRI and other institutions around the country. In several courses foreign faculty members were associated.

There was great interest at IVRI in this new area of biology which had tremendous application in diagnostics/vaccines/reproductive biotechnology/genetic improvement of livestock. British council supported two major projects under their programe in (i) Recombinent DNA technology and its application for producing recombinant vaccines, and (ii) embryo transfer technology and associated reproduction technologies. All had emphasis on human resource development and publication of workable monographs.

The present project with Studium Press (India) Pvt. Ltd. publication is the culmination of the idea of Dr S. Ramachandran in order to develop text books based on Indian experience, which has made a remarkable progress. Prof. Pran P. Bhat was the group leader of a group of young scientists, students and faculty who joined hands with support from undersigned to convert these monographs into as the texts for the students. This new Scientific endeavor has had support from DBT and the publication house.

In tune with this thinking, it was decided to bring out a series of books for under graduate and post graduate scholars in Biosciences, Animal, Veterinary Sciences and Agriculture in seven (7) volumes under the Chief editorship of Professor (Dr) P.N. Bhat, Former Vice Chancellor and Director of Indian Veterinary Research Institute at Izatnagar 243 122 (U.P) and Former Animal Husbandry Commissioner of India and Deputy Director-General (Animal Science) of Indian Council of Agricultural Research, Ministry of Agriculture, Krishi Bhawan, New Delhi. The basic idea of the series is to provide first rate text books to students and scholars in developing countries based on the experiences of India with special focus on Indo-Pak sub-continent in conformity with standards laid out by regulatory agencies in India (VCI, ICAR, UGC, AICTE, IMC, CSIR) and similar agencies in other developing countries.

The series includes books on

1. Animal Cell Culture
2. Hybridoma Technology
3. Cell and Molecular Biology
4. Genetic Engineering
5. Microbial Fermentation
6. Feed Biotechnology
7. Embryo Biotechnology

It is hoped that in next 2–3 years these books will be available for the benefit of students, teachers and professionals in this area and fill the gap which is becoming wide.

Prof (Dr) Pushkar Nath Bhat
Chairman- World Buffalo Trust
and
Centre for Integrated Animal Husbandry & Dairy Development
and
Chief Editor

ABOUT THE AUTHORS

Prof (Dr) P.N. Bhat got his Ph.D in population genetics from Institute of Population Genetics Purdue University, West Laffayette, Indiana, USA followed by several post doctoral assignments and visiting professorship in genetics, biotechnology, livestock production systems. He returned to India and joined Punjab Agricultural University at Hisar–Ludhiana followed by several professional assignments in India and abroad. He joined as Project co-ordinator (Animal Breeding) and subsequently as Head, Division of Animal Genetics at Indian Veterinary Research Institute in 1971. He was responsible for establishing coordinated projects on cattle, buffalo, sheep, goat, pigs and poultry during 1970–74. He was founder Director of Central Institute for Research on Goats. In 1984 he became Vice-Chancellor and Director of IVRI. He joined as Deputy Director General (Animal Science) in May, 1992 and Animal Husbandry Commissioner in December, 1992. He is fellow of several National and International Science Academies.

Dr. P.P. Bhat got her Ph.D in Animal Genetics in 1971. She joined as Assistant Professor Animal Genetics, Scientist S3 at CIRG, Course coordinator and Principal Investigator- Animal Biotechnology, educational programme of Department of Biotechnology and established National Biotechnology Centre at IVRI, Principal Scientist (Education Division, ICAR), Principal Scientist (AG&B), Animal Science Division ICAR, Assistant Director General (AP &B) ICAR. She was receipient of PAU Fellowship Award 1966-70, Lal Bahadur Shastri Award for excellence in Science (Biotechnology) by Prime Minister 1988, Dr. R. Eswaran Memorial Medal 1993, Award of gold Record of Achievement from American Biographical Institute Inc. USA 1996. Kashmir Education, Cultural and Science Society Award 2003. She is Fellow of New York Academy of Science, National Academy of Agricultural Sciences and ICAR National Fellowship. She is associated with Animal Genetics and Breeding, Animal Biotechnology (Production and Health).

Dr. B.P. Mishra got his Ph.D. in Animal Genetics (Molecular Genetics) from National Research Centre for Biotechnology, Indian Veterinary Research Institute, Izatnagar in the year 1996 followed by a post doctoral fellowship in biotechnology in Department of Animal Science, Iowa State University, Ames, Iowa, USA. He attended several short term training programmes in Molecular Biology, Biotechnology and Marker Assisted Selection (MAS). He joined National Research Centre for Biotechnology as Technical Officer dealing with recombinant DNA project in 1989. He joined as Senior Scientist in Animal Genetics and Breeding at National Research Centre on Camel at Bikaner in 1996, in the year 2000 he joined at National Bureau of Animal Genetics Resources, Karnal as Senior Scientist and currently he is Principal Scientist, Animal Biotechnology. In the past 20 years he worked in the general area of Molecular Genetics and Animal Biotechnology and has been associated with different finding with regard to Gene Expression, sequencing in Buffalo Mammary Gland. He is also been responsible for identification of QTL'S for dwarfing gene in Angus cattle. He is currently involved in gene mining of livestock resources.

Dr. (Mrs.) Bina Mishra got her Ph.D. in Microbiology and subsequently worked for 12 years in Animal Biotechnology and Microbiology. She is currently engaged in vaccine development. She is currently working as Senior Scientist at Indian Veterinary Research Institute, Izatnagar on Vaccine development using Biotechnology tools.

PREFACE

This book on Animal Cell culture is intended as an introductory book for those who are new to cell culture, whether undergraduates, graduates, technicians or scientists who want to move into the general area of biotechnology.

As researchers who have both spent much of their time at the bench, performing cell culture, we recognize the need for an inexpensive, introductory text for the first timers who study biotechnology for research or as a profession. Many comprehensive works are already available in this field, but the sheer number and variety of current techniques in animal cell culture has resulted in long and expensive texts. This book is not meant to give a comprehensive list of every available protocol; it aims to explain why and how the basic techniques are used, and their applications in modern molecular biology and biotechnology.

The initial chapters describe the basic methodologies needed for the culture of cells, including laboratory design, sterilization techniques, media preparation, and guidelines for the correct use of equipment. The basics of performing cell culture on primary as well as established cell lines are described, together with the essentials of dealing with contamination. The storing of cells by cryopreservation completes the know how of basic and essential techniques. Armed with this necessary background, these techniques can be applied to the cloning of cells, the separation of lymphocytes from various sources and the use of these in cell fusion, cytotoxicity assays and the immortalization of various cells.

It is hoped that this book may act as a stimulus for further, more detailed, reading. Theoretical considerations have been included where necessary for a better understanding of the techniques described. Much of what is contained in this book is the result of the accumulated knowledge and experience of countless researchers who have refined techniques to the point where they can give reliable results even in the hands of inexperienced workers. We hope that with the feed–backs from the students/scientist who use this text, the subsequent editions will be reorganized based on their experiments.

New Delhi
20th January, 2011

Bina Mishra
B.P. Mishra
Pran P. Bhat
P. N. Bhat

CONTENTS

LIST OF FIGURES

LIST OF TABLES

CHAPTER 1

INTRODUCTION

1.1 Introduction

Cell culture is the technique of growing and maintaining cells of multi-cellular organisms *in vitro* conditions which mimic the precise environmental conditions such as temperature, humidity, nutrition, and contamination-free conditions that were present *in vivo*.

The term 'cell culture' refers mainly to the culturing of cells from animals and humans, namely eukaryotic cells. The different cell types which can be grown in culture include connective tissue elements such as fibroblasts, skeletal tissue (bone and cartilage), cardiac and smooth muscle, epithelial tissue (liver, lung, skin, bladder and kidney), neural cells, endocrine cells (adrenal, pituitary, pancreatic islet cells), melanocytes and many different types of tumor cells.

Animal cells selected for culture can be either identical (homogenous population) or may show genetic variation (heterogeneous population). The culture of individual identical cells is the preferred technique because conditions can be controlled to allow same degree of consistency and reproducibility in cell growth.

A variety of areas of research in biotechnology depend largely on availability of technique of culturing of tissues and cells and therefore, the development of tissue culture methods in animals have received great attention. These areas include: (i) production of antiviral vaccines, which requires the standardization of cell lines for the multiplication and assay of viruses; (ii) cancer research, which requires the study of uncontrolled cell division in cultures; (iii) cell fusion techniques; (iv) genetic manipulation, which is possible in cells or organs in culture; (v) monoclonal antibodies whose production, requires cell lines in culture; (vi) production of pharmaceutical drugs using cell lines; (vii) chromosome analysis of cells; (viii) study of the effects of toxins and pollutants using cell lines; and (ix) study of the function of cell types.

1.1.1 Importance of cell culture

Cell culture has become one of the major tools used in cell and molecular biology. Some of the important areas where cell culture technology is currently playing a major role is briefly described below:

(a) *As model systems*: Cell cultures provide a good model system for studying:

1. Basic cell biology and chemistry, which enables investigation of the normal physiology or chemistry of cells *e.g*. metabolic pathways
2. The interactions between disease causing agents and cells.
3. Effects of drugs on cells.
4. To elucidate the effects of chemical compounds, metabolites, hormones or growth factors on specific cell types.
5. To evaluate potentially toxic or mutagenic compounds in cell culture.
6. Study the process of aging.
7. Nutritional interactions.
8. To produce artificial tissue by combining specific cell types in sequence.

(b) *Toxicity testing*: Cultured cells are widely used alone or in conjunction with animal tests to study the effects of new drugs, cosmetics and chemicals.

(c) *Cancer research*: Since both normal and cancer cells can be grown in culture, the basic differences between them can be closely studied. It is possible, to find methods/agents (chemicals, viruses and radiation), to convert normal cultured cells to cancer causing cells. Thus, the mechanisms that cause the change by which they lead to uncontrolled prolificacy can be studied. Cultured cancer cells also serve as a test system to find suitable drugs and methods for treating several types of cancer.

(d) *Virology*: One of the earliest and major uses of cell culture was in the replication of viruses in cell cultures (in place of live animals) for use in vaccine production. Cell cultures are widely used in the clinical detection and isolation of viruses. Basic research in virology to understand pathobilogy of viruses in *in vitro* cell culture.

(e) *Cell-based manufacturing*: Cells are used to synthesize valuable products (biologicals) from large scale cell cultures. (1) The large-scale production of viruses for use in vaccine production, which include vaccines for polio, rabies, chicken pox, hepatitis B, FMD, measles and many more. (2) The large-scale production of cells that have been genetically engineered to produce proteins that have medicinal and commercial value. These include (i) monoclonal antibodies, (ii) insulin, (iii) hormones, etc. (3) The use of cells for replacing tissues and organs *e.g.* artificial skin for use in treating burns. A potential supply of replacement cells and tissues may come from embryonic and adult stem cells. These are cells that have the potential to differentiate into a variety of different cell types.

(f) *Genetic counselling*: Amniocentesis, a diagnostic technique that enables doctors to remove and culture fetal cells for the early diagnosis of fetal disorders. These cells can then be examined for abnormalities in their chromosomes and genes using karyotyping, chromosome painting and other molecular techniques to identify genetic disorders.

(g) *Genetic engineering*: The ability to transfect cultured cells with new genetic material (DNA and genes) has provided a major tool to study gene expression and further can be used for bulk production of specific proteins in cultured cells.

(h) *Gene therapy*: The ability to genetically engineer cells has also led to their use for gene therapy. Cells can be removed from a patient, lacking a functional gene and this missing or damaged gene can then be replaced, by cells grown for a while in culture and then replaced in the patient. An alternative approach is viral vector mediated gene transfer in patient with the hope that the missing gene will then be expressed in the patient's cells.

(i) *Drug screening and development*: Cell-based assays have become increasingly important for the pharmaceutical industry, not just for cytotoxicity testing but also for high throughput screening of compounds that may have potential use as drugs. Originally, these cell culture tests were done in 96 well plates, but increasing use is now being made of 384 and 1536 well plates.

(j) *Stem cell research*: A lot of progress has been also made in the area of stem cell technology which will have their use in the possible replacement of damaged and dead cells. In 1997, Wilmut and co-workers successfully produced a transgenic sheep named Dolly through nuclear transfer technique. Thereafter, many such animals (like sheep, goat, pigs, fishes, birds etc.) were produced.

(k) *Cell based therapy*: The animal cell culture techniques are used in replacing the damaged and dead cells with normal and healthy cells using the stem cell technology. This therapy is called cell-based therapy which involves the use of stem cell technology involving the replacement of damaged and dead cells with normal and healthy cells. This is used to treat blood cancer, and other neuro-degenerative diseases etc.

1.2 Historical Development

The technique of culture of cells *in vitro* is one of the great discoveries in the field of biology which has transformed research and development portfolio in the fields where biological sciences are required be it medicine, agriculture or any other aspect of human endeavour including energy. Initial reports of tissue culture activities began to appear in scientific literature in the mid 1880's. Most often cited work was related to the maintenance of chick embryonic tissue in warm saline (Roux, 1885). Researchers demonstrated that cultured cells could survive, move, and proliferate *in vitro* under "appropriate conditions".

The demonstration of coagulated plasma (Burrows, 1910) led to expansion of culture application possibilities. The clot technique was a valuable early research tool that generated considerable interest. It was later recognized that the majority of cultured cells do not require the supporting matrix that was provided by clot (lymph/plasma) techniques. The science of tissue culture had acquired a new critical dimension: analysis of basic constituents necessary for the propagation of cells on matrix. It lead to define the nutritional requirements for individual *in vitro* culture application(s). During the first five decades of the present century tissue culture met with occasional culture successes in supplemented, buffered saline solutions or plasma clots.

Another significant advance was the use of trypsin (a proteolytic enzyme) by Rous and Jones in 1916 to free cells from tissue matrix. It was subsequently used for the subculture of adherent cells and in the 1950s, the technique of trypsinization was exploited to produce homogeneous cell strains and this marked the start of animal cell culture techniques. Trypsinization is the term applied to the treatment of cells by the proteolytic enzyme trypsin to change their adhesiveness. The term '*cell culture*' refers to cultures derived from dispersed cells taken from the original tissue. In cell culture the cells are no longer organized into tissues.

Many animal cell types are known to have the capacity to survive, multiply and differentiate under "appropriate" *in vitro* conditions. The extra cellular medium must supply all essential nutrients. Controlled component experiments began in the early 1900s with the examination of the effects of sea water, serum, embryo extracts, and peptones at varying concentrations (Lews and Lewis, 1911). It took another 20 years when importance of amino acids in growth medium was described (Willmer and Kendall, 1932). Attempts to develop synthetic medium continued during the 1940s when dialyzed plasma was used as a basal medium and supplemented with amino acids similar to those found in fibrin. Eagle's basal medium formulation, characterized as an osmotic, pH-balanced mixture of salts, amino acids, sugars, vitamins and other necessary nutrients, was described in 1955. Such media were considered complex by formulation but not complete without the addition of a protein source such as plasma, serum, body fluids or tissue extract. The importance of these medium formulations and their success act as models for the evolution and development of most common media those exists today.

Milestones in cell culture	
Claude Bernard, (1878)	Proposed that physiological systems of an organism can be maintained in a living system after the death of an organism.
Wilhelm Roux, (1885)	First recorded instance of explanation of modularly flat of a chick embryo in warm saline.
Arnold, J. (1887)	Survival and migratory capability of frog leucocytes in saline.
Junggren, L. (1898)	Reimplantation human skin could survive *in vitro* for many days when stored in ascetic fluid.
Tyrode, (1910)	First salt solution specially for mammalian cells proposed.
Carrel and Burrows, (1911)	Tissue extracts promote *in vitro* growth of cells.
Warren and Margaret Lewis, (1911–1912)	Development of common media.
Carrel, A. (1913)	Direct relationship between the application of asepsis and successful culture technique.
Ebeling, (1913)	First to apply tissue culture to warm-blooded animals. They grew chick embryo tissue in fowl plasma/serum by hanging drop method.
Carrel Alexis, (1913)	Embryo extract-strong growth promoting effect. Demonstrated without doubt that animal cells grow indefinitely *in vitro*.
Rous and Jones, (1916)	Introduced trypsinization.
Willmer and Kendall, (1932)	Importance of amino acids in growth medium described.
Kurotchkin, (1939)	Modifications of technique cultivation in agar slant cultures.
Willmer, (1945)	Demonstration of dispersive effect of trypsin on embryonic-chick heat cells.
Moscona and Moscona, (1952)	Used trypsin as a means of dispersing viable cells from their parent tissues/organs.
Baker with carrel, Fischer, Parker, Healy, Morgan, White, (1950); Earle, (1950); Waymouth and Eagle, (1957)	Development of present day medium (composition and factors of the medium).

Table (*Contd...*)

Table (*Contd...*)

Milestones in cell culture	
Dulbecco, (1952)	Used trypsin dispersal method of Rous and Jones for the dispersion of tissue fragments into cell suspension.
Earle's group, (1954)	First attempt to grow fibroblastic cells into suspension at National Cancer Institute, USA.
Hayflick and Moorhead (1961)	Finite life span of normal human cell.
Ham, (1965)	Cloning of Chinese hamster cell with out using serum.
Kohler and Milstein, (1975)	Developed plaque assay for animal viruses using confluent monolayer of cultured cells.
Thomson *et al*, (1998)	Culturing embryonic stem cells of human.

1.3 Land Mark Achievements

Although it would seem logical to begin with a description of cell culture and then proceed via tissue culture to organ culture, historically it was tissue culture that developed first. Tissue culture paved the way for the later development of cell and organ culture. Tissue culture, as it is practiced today, can be considered under three headings. First the "tissue culture" proper, in which small fragments of tissue are explanted into a suitable medium and encouraged to grow in isolation, to form colonies, and perhaps to continue some of their functions. In such cultures, the original organization of the tissues may be lost, but the constituent cells emerge into the outgrowth where their activities can be directly observed. Secondly in "cell culture" the cells of a tissue, or even individual cells, are made to grow. The organization of the original tissues in such cultures is irrelevant but cell multiplication and growth in uniform populations are of main interest. The cells of different tissues tend to become undifferentiated and relatively alike in appearance. In organ culture, growth is only of minor interest, but embryological development and the maintenance of normal physiological functions are the main aim and object. The outward growth and migration of "undifferentiated" cells is positively suppressed and the maintenance of the normal organization (architecture) of the tissue is of primary interest. The development of watch-glass technique (Fell and Robinson, 1929) marked the true beginning of organ culture as opposed to tissue culture and cell culture.

1.3.1 Tissue culture

Tissue culture is often a generic term that refers to both organ culture and cell culture and the terms are often used interchangeably. The culture may be obtained from tissue fragment having heterogeneous population of cells. This can be propagated but up to some extent only. Characterization on the basis of cytological markers is possible. Originally, the term is not limited to animal cells, but includes the *in vitro* cultivation of plant cells also. Tissue culture can be subdivided into three major categories; organ culture, explant culture, and cell culture.

1.3.2 Cell culture

Features of cell culture are development of a finite cell line over several generations from untransformed tissue. Monolayers obtained may be sub cultured enzymatically into fresh flask. This constitutes a passage and the daughter cultures so obtained gives rise to a cell line. Characterization of these cell lines can be done by cytological, biochemical, immunological and molecular assays. Cells may be propagated following standard procedure of sub-culturing to attain uniform cell population. The cell line can be grown in monolayer or suspension culture.

1.3.3 Organ culture

An organ culture is *in vitro* culture and growth of organs or parts of organs there of in which their various tissue components, *e.g.* parenchyma and stroma, are preserved both in terms of their structure and functions so that the culture organs resemble closely the concerned organs *in vivo* is called organ culture. The organ culture has the following advantages.

1. The explants remain comparable to the *in vivo* organs both in structure and function, which makes them more suitable than cell cultures for physiological studies.
2. The development of fetal organs *in vitro* is comparable to that *in vivo*.
3. Organ cultures provide information on the patterns of growth, differentiation and development, and the influences of various factors on these features.
4. In some cases, organ cultures may replace whole animals in experimentation as the results from them are easier to interpret.

However, organ culture suffers from the following limitations.

1. Results from organ cultures are often not comparable to those from whole animal studies, *e.g.* studies on drug action, since the drugs are metabolized *in vivo* but not *in vitro*.
2. Organ cultures can be maintained only for a few months. But it may be desirable to study the effects of certain factors for several months. In such cases, the organs treated *in vitro* may be transplanted into suitable host animals, *e.g.* nude mice.

1.4 Cells in Culture

The proliferation of animal cells in culture has many applications, including their use as model systems for biochemical, physiological and pharmacological studies, and to the production of growth factors, blood factors, monoclonal antibodies, interferons, enzymes, vaccines and hormones. Advances in animal cell culture applications have been profound in recent years because of the versatility of the techniques and the ready availability of media, cells and culture vessels from specialist suppliers. The major drawback is that such techniques may be expensive to operate and require specialized training.

Examples of cells in culture include macrophages, T and B lymphocytes and polymorph nuclear leukocytes. Pioneering studies established that varying degree of growth of tissue was possible *in vitro* in sterile medium containing salts, carbohydrates, vitamin, amino acids and serum supplements. Animal cells may be subdivided broadly into those that remain viable only when attached to a solid substrate (*e.g.* primary cultures or diploid fibroblast cells) and those that will proliferate as suspensions (*e.g.* Human tumor cells, HeLa cells). Newly isolated or primary cell cultures of non-tumor cells typically show anchorage dependence by growing *in vitro* only when attached to a surface, which may be either other cells or glass, plastic, gelatin or collagen. Primary cell lines spread over the surface as monolayer but will not overlay due to contact inhibition of adjacent cell membranes. Such cultures are said to show confluence, which severely limits cell density in culture.

Continuous subculture is possible only in transformed cell lines that have become naturally aneuploid or chemically mutated (*e.g.* 3–4 benzpyrene, nitrosomethyl urea) or infected with a virus, *e.g.* simian virus 40 or Rous sarcoma virus. Transformed cells, *e.g.* Baby Hamster Kidney (BHK) cells typically show reduced contact inhibition, an ability to grow in suspension, changes in shape and loss of differentiation and a requirement for much less stringently defined culture medium. Interestingly, some transforming viruses may introduce viral oncogenes that code for proteins substituting for growth factors that are normally supplied in serum. Transformed cell lines frequently induce tumor formation when injected into suitable hosts. The close similarity between transformed and tumor cells is often exploited in experiments, with transformed cells being used instead of tumor cell because of their tendency to be cytologically more homogenous.

Small-scale static culture of cells up to 1cm^3 is conveniently established either in microtitre plate wells or specialized tissue culture plates having wells, of varying cubic capacity. Specialized, and expensive, cell factories or capillary perfusion beds are also available, which employ either glass bead columns or hollow fibers to support cells, in which medium previously equilibrated with 5% CO_2 in air is circulated.

Suspension culture of lymphocytes or hybrid cells, for example, is possible in small-scale fermenters of up to 10dm^3 capacity, in which cells are magnetically stirred gently. Alternatively, small-scale air lift fermenters may be employed. Free cell suspension culture has very limited application, while development of micro carriers, allows for large-scale suspension culture of anchorage-dependent cells. Micro carriers consists of a cross-linked polymer, such as polyacrylamide, polystyrene or plastic, usually containing either fixed positive or fixed negative charges that facilitate attachment.

1.4.1 The characteristics of animal cells in culture

A primary cell culture is established by inoculating cells taken directly from animal or human tissue into growth medium. The tissue is first fragmented into small pieces which are placed into a sterile medium in a petri-dish. Treatment of the excised fragments with a proteolytic enzyme, such as trypsin disaggregates the tissue into individual cells.

Cell types may be recognized by their characteristic shapes when examined under a microscope. Fibroblasts which are spherical upon treatment with trypsin, elongate to a spindle shape on attachment to a surface. Epithelial cells, which also have a requirement for surface attachment, have a characteristic cobble stone appearance.

Most cells derived from animal tissues require a substrate for growth and are termed anchorage-dependent. However, cells associated with biological fluids, *e.g.* blood cells (lymphocytes) are non-anchorage-dependent and may be grown in suspension .

A secondary culture is established by sub culturing a primary culture. In the case of suspension cultures, such as lymphocytes, this involves diluting the culture when the cells have grown to a high density with fresh growth medium. Anchorage-dependent cells from a monolayer which eventually may completely cover the surface on which they are growing termed confluence. Sub-culturing involves the removal of cells (by trypsinization) and their re-inoculation into fresh medium. The cells may grow at a constant rate over successive subcultures.

Hayflick and Moorhead (1961) studied the growth of cells derived from normal human embryonic tissue and showed that these cells could be grown continuously through repeated subculture for about fifty generations. After this period the cells enter a senescent phase in which there was an observed decline in cell numbers. A finite growth capacity of this type is characteristic of all cells derived from mammalian tissues. The extent of growth is related to the origin of the cells like those derived from embryonic tissue have a greater growth capacity than those derived from adult tissue.

1.4.2 Behaviour of cells in culture

Epithelial cells, from the covering layers of a wide variety of ectodermal, endodermal and mesodermal tissues, emerge as sheets of cells closely adherent to each other. Fibroblasts from a wide variety of sources are a loose network of cells, in which the cells are polarized and partly contiguous. Amoebocytes (macrophages), derived from monocytes in the blood, show much more random movement, and these cells remain isolated from each other. These also may, however under some conditions produce loosely knit sheets of cells. They also have a tendency to aggregate into giant cells. Schwan cells may emerge as rather characteristicaly spindle-shaped cells or as cells appearing more like amoebocytes (macrophages). Lymphocytes on a flat surface show either the "hand-mirror" type of gliding movement or they round up and only produce pseudopodia. Lymphocytes, like all the other types of blood cells except the monocytes, only survive for a few hours or days *in vitro* and then disintegrate.

1.4.3 Differentiation

As propagation of cell lines requires that the cell number increases continually, culture conditions which have evolved over the years have been selected to favour maximal cell proliferation. It is not surprising that these conditions are not often conducive to cell differentiation where cell growth is severely limited or completely abolished. Those conditions which favour cell proliferation are low cell density, low Ca^{2+} concentration (100–600 uM) and the presence of growth factor such as epidermal growth factor (EGF), fibroblast growth factor (FGF), and platelet derived growth factor (PDGF). High cell density (> 10^5 cells/cm^2), high Ca^{2+} concentration (300–1500 μM) and the presence of differentiation inducers (hormones such as hydrocortisone, glia maturation factor, nerve growth factor, dimethy-1 sulphoxide etc.), will favour cytostasis and differentiation. The role of serum in differentiation is not entirely clear and depends on the cell type and medium used. While low serum concentration promotes differentiation in oligo-dendrocytes, a high serum concentration causes squamous differentiation in bronchial epithelium.

CHAPTER 2

LAB DESIGN, EQUIPMENTS AND MATERIALS

The major pre-requisite for a tissue culture laboratory is the need to maintain asepsis. It is important therefore that the tissue culture laboratory be dust free, and have no through traffic. People require access to the animal facility for animal tissue, but it must be ensured that tissue culture laboratory is reasonably accessible to, but not contiguous with, the animal facility. Thus, several considerations need to be taken into account, in planning tissue culture labs. The location needs to be choosen carefully and it should be assured that doorways are both wide and high enough and ceilings have sufficient clearance to allow the installation of equipment such as laminar-flow hoods, incubators, and autoclaves. It is essential that doorways and spacing between equipments provide easy access for maintenance.

2.1 Design and Layout

(a) ***Floor plan*****:** The tissue culture laboratory should be laid out in such a manner as to provide maximum protection for open cultures from contamination. This is achieved by designing the laboratory to provide, environmental control around the cultures. The basic layout of most tissue culture laboratories has a Laminar Flow Cabinet (LFC) in the center, which is the sole barrier to contamination, surrounded by a room or rooms that are totally without environmental control. Fig. 2.1 describes the ideal laboratory layout. In the ideal case the LFC is contained in an environmentally controlled room. Access to this room is through another environmentally controlled area, but one of less stringent air quality.

The areas where cultures are open to the environment are primary areas. Rooms containing the primary areas are secondary areas. Ancillary areas are all other controlled areas that support the secondary areas *viz.* component preparation rooms, air locks, and change rooms etc.

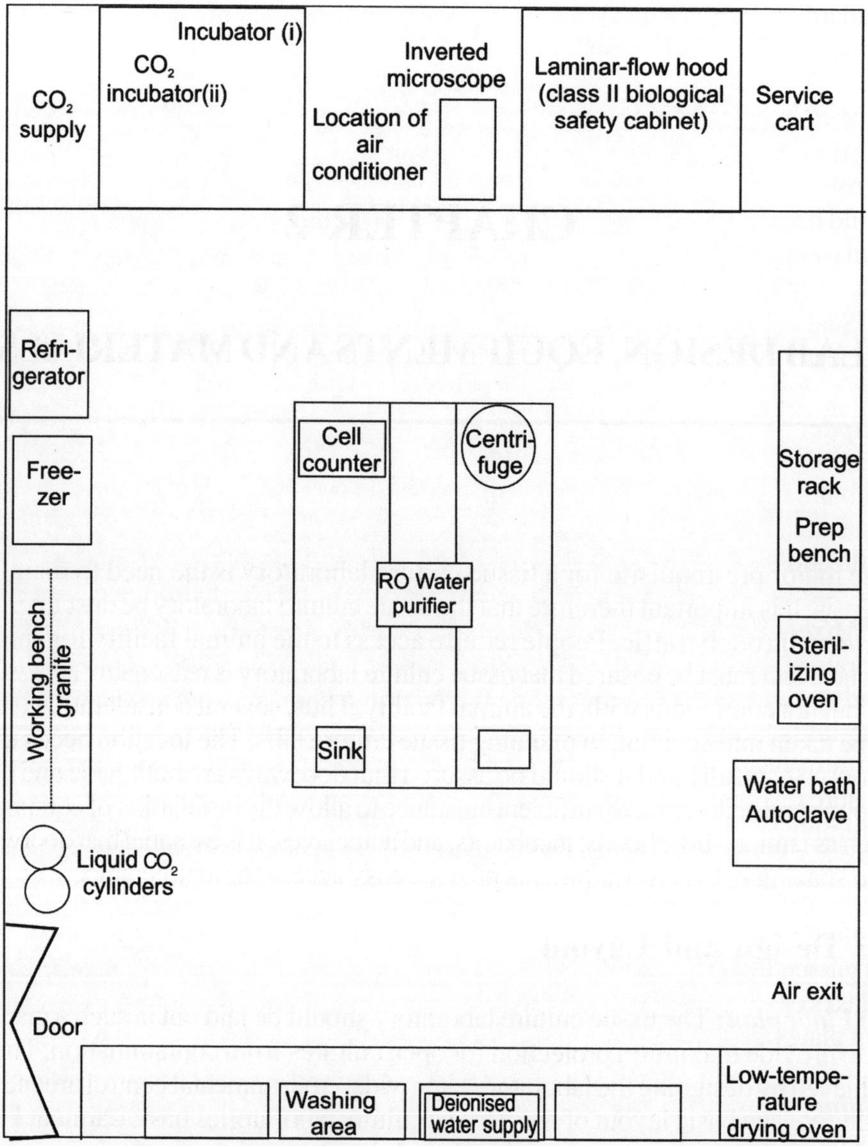

Fig. 2.1. Basic layout for simple, self-contained tissue culture laboratory for use by two or three persons

Temperature and humidity control generally should be at a temperature range of 25° to 27°C which would be comfortable working temperature and also tend to shed less microorganism that at warmer temperature.

A written standard operating procedure (SOP) should exist for monitoring the laboratory to determine that it is being operated within the written specifications.

Standard operating procedures (SOP) should define the frequency of monitoring, the monitoring sites, and what action is to be taken.

(b) *Walls and ceilings*: Walls and ceilings should be with coated surfaces, sealed at the joints to eliminate any crevices that could hold dust particles or microbes. Floors should be nonporous and seamless. Shelving if it is necessary should be confined to the ancillary areas and should be of the type designed for clean room use.

Seal all openings around the fixtures to avoid dust. It is desirable that all water and electricity fittings should be concealed in walls.

(c) *Space*: The largest area of tissue culture lab should be given to the culture operation, which has to accommodate laminar-flow hoods, cell counters, centrifuges, incubators, microscopes, refrigerators and some stocks of reagents, media, glassware, and plastic wares. The second largest is for wash-up, preparation, and sterilization, third is storage, and fourth is incubation.

(d) *Miscellaneous aspects of design*: *Pressure balance*: A tissue culture laboratory should be at positive pressure relative to surrounding work areas, to avoid any influx of contaminated air from outside. However, if biohazard pathogen material is being used, most safety regulations require that the tissue culture laboratory be designated as Category–I and this requires that it is at negative pressure relative to the surrounding areas. To satisfy both requirements it may be preferable to have a positive-pressure buffer zone outside the tissue culture laboratory, such as the Preparation Area and Microscope Room or the corridor.

***Incubation*:** Type of incubation required in terms of size, temperature, gas phase, non gassed incubators and proximity to the work space. Generally, large number of flasks or large-volume flasks that are sealed are incubated in a hot room, while open plates and dishes will require a humid CO_2 incubator.

***Heating, ventilation and air conditioning (HVAC)*:** The HVAC system should be dedicated to the tissue culture area with no return air coming from other areas.

***Quarantine*:** Newly introduced cell lines need to be screened for mycoplasma before being handled in the same room as general stocks, and some human and primate biopsies and cell lines may carry a biohazard risk that requires containment. If possible it is preferable for the tissue culture lab to be separated from the preparation, wash-up, and sterilization areas, while still remaining adjacent.

***Water*:** Floor drains should be sealed and sinks removed. These are both potential sources for microbes. Water baths should be filled with a mixture of water and disinfectant, in the appropriate proportions this mixture should be replaced on a regular basis.

2.2 Construction and Services

The rooms should be supplied with filtered air. If Category-I containment is required for the tissue culture laboratory it will be required to be at negative pressure relative to other areas. In this case adjacent rooms, which lead to the aseptic area, should be regarded as buffer zones and should also receive filtered air but at positive pressure.

The rooms should be designed for easy cleaning. The floor is covered with a vinyl or other dust proof finish, and allow a slight fall in the level toward a floor drain located outside the door of the room (*i.e.*, well away from the sterile cabinets). If possible it is preferable for the tissue culture lab to be separated from the preparation, wash-up, and sterilization areas, while still remaining adjacent. Adequate floor drainage should be provided in the preparation/wash-up area, with a slight fall in floor level from the tissue culture lab to the wash-up. The lab should be designed and arranged in such a way that there is minimum conflict, easy and close access, less interfering with sterile work, and also easy removal of discarded items.

A vacuum line is generally preferred removing contents from culture flasks. While using vacuum line, a collection vessel must be kept with an additional trap flask, with a hydrophobic filter between the flasks, in order to prevent fluid, vapor, or some contaminant from entering the vacuum line and pump.

2.3 Layout of Aseptic Room

The important functions to be accompanied in the tissue culture lab are: sterile handling area; incubation; preparation area; washing area; sterilization area; and storage area. The sterile handling should be located at one end of the tissue culture lab, farthest from the washing and sterilization area which are placed at the other end. The preparation, storage, and incubation areas should be located in between. The preparation area should be adjacent to the washing and sterilization areas, while incubators should be easily accessible to the sterile working area.

Sterile work should be restricted to one part of the tissue culture laboratory without through traffic or other disturbance that is likely to cause dust or drafts. Most cell culture handling is preferably carried out in Class II Biological Safety Cabinets (BSC). These cabinets maintain a clean working environment for cell handling and help to provide protection to the operator and environment. Horizontal laminar flow hoods are useful for media preparation but are not desirable for cell culture work. The effectiveness of a BSC/laminar flow hood is dependent on its location in the room, regular testing and maintenance.

While installing laminar flow hood in a room, consider where air inlets and exits must be placed. It is preferable to duct laminar-flow hoods to the exterior to improve air circulation and remove excess heat from the room. This also facilitates decontamination with formaldehyde.

2.3.1 Quarantine and containment

If sufficient space is available, it is worth designating a separate room as a quarantine and/or containment room. This is a separate aseptic room with its own laminar-flow hood (Class II microbiological safety cabinet), incubators, freezer, refrigerator, centrifuge, supplies, and disposal. This room must be separated by a door or air locks from the rest of the suite and be at negative pressure to the rest of the aseptic area. Newly imported cell lines or biopsies can be handled here until they are shown to be free of contamination, particularly mycoplasma, and pathogens etc. If used at a higher level of containment it will also require a biohazard cabinet or pathogen hood with separate extract and pathogen trap.

2.3.2 Incubators

Non-CO_2 standard incubators are simple, effective and usually reliable which require all cell culture vessels to be sealed in order to maintain proper pH for cell culture (between 7.2 and 7.4) as well as prevent evaporation of the culture medium. All cell culture flasks, tubes or single/muti-well TC plates in a non-CO_2 incubator must be well sealed. Seal the lids of microtitre plates with non-toxic sealing tape or place them in a sealed plastic box with moistened paper in the bottom. CO_2 incubators provide humidity and a 5% CO_2 atmosphere for cell culture.

2.3.3 Preparation area

The preparation area in a cell culture lab should be large enough to accommodate a variety of equipments needed for general purpose along with enough bench space for making solutions, media and reagents etc. Preparation and packaging various materials, laminar flow hood for filtering and bottling sterile liquids, and separate incubator space for quality control of sterility assay.

2.3.4 Sterilization

Facilities for washing and sterilization should be located close to the aseptic area so that washing and sterilization should be checked regularly. The size and number of the autoclaves and hot air oven depends on the work load in the lab. They should not be overloaded while in use. Regular maintenance for the autoclaves and hot air ovens are required for the effective and accurate results of the lab.

2.4 Equipments

A good tissue culture lab depends on how well equipped the lab is? Where possible, a separate room should be made available for clean cell culture work, equipped with an air flow cabinet which supplies filtered air around the work surface. Clean laboratory coats should be kept at the entrance and should not be worn outside of this laboratory and brought back in. The basic minimum or essential equipments required for cell culture work are: (i) Laminar flow hood(s); (ii) Incubator; (iii) CO_2 cylinder; (iv) Balance (coarse and fine); (v) Sterilizer (autoclave, pressure cooker, oven); (vi) Refrigerator; (vii) Freezer (for –20°C storage); (viii) Inverted microscope; (ix) Soaking bath or sink; (x) Deep washing sink; (xi) Pipette cylinder (s); (xii) Pipette washer; (xiii) Water purifier; (xiv) Bench centrifuge; (xv) Deep freezer (–70°C) and (xvi) Liquid N_2 storage tanks.

2.4.1 Laminar flow hoods

There are two types of laminar flow hoods, vertical and horizontal. The vertical hood, also known as a biological safety cabinet, is best for working with hazardous organisms since the aerosols that are generated in the hood are filtered out before they are released into the surrounding environment. Horizontal hoods are designed such that the air flows directly at the operator hence they are not useful for working with hazardous organisms but are the best protection for your cultures. Both types of hoods have continuous displacement of air that passes through a HEPA (high efficiency particle) filter that removes particulates from the air. In a vertical hood, the filtered air blows down from the top of the cabinet; in a horizontal hood, the filtered air blows out at the operator in a horizontal fashion. The hoods are equipped with a short-wave UV light that can be turned on for a few minutes to sterilize the surfaces of the hood, but be aware that only exposed surfaces will be accessible to the UV light. Do not put your hands or face near the hood when the UV light is on as the short wave light can cause skin and eye damage. The hoods should be turned on about 10–20 minutes before being used. Wipe down all surfaces with ethanol before and after each use. Keep the hood as free of clutter as possible because this will interfere with the laminar flow air pattern.

2.4.2 Incubator

Usually for most mammalian cells 37°C temperature with humidity 95% and 5% CO_2 tension is required, whereas non mammalian cells might require different culture conditions. Depending on the need of the cell culture work selection of the incubators is done.

Standard incubators (non-CO_2)

These incubators are simple, effective and usually very reliable. All cell culture flasks, tubes or plates in a non-CO_2 incubator must be well sealed in order to maintain proper pH for cell culture (between 7.2 and 7.4) as well as prevent evaporation of the culture medium. Close the lids of flasks or tubes securely. Seal the lids of microtitre plates with non-toxic sealing tape or place them in a sealed plastic box with moistened paper in the bottom. This helps to reduce the risk of cross-contamination with other cultures.

CO_2 incubators

The cells are grown in an atmosphere of 5–10% CO_2 because the medium used is buffered with sodium bicarbonate/carbonic acid and the pH must be strictly maintained. Culture flasks should have loosened caps to allow for sufficient gas exchange. Cells should be left out of the incubator for as little time as possible and the incubator doors should not be opened for very long. The humidity must also be maintained for those cells growing in tissue culture dishes so a pan of water is kept filled at all times. The incubation temperature will depend on the type of cells being cultivated. Insect cells will grow best at around 30°C while mammalian cells require a temperature of 37°C. It may be necessary to use an incubator, which has been designed to allow CO_2 to be supplied from a main supply or gas cylinder so that an atmosphere of between 2–5% CO_2 is maintained in the incubator. In general, many cell lines can be maintained in an atmosphere of 5% CO_2, 95% air at 99% relative humidity. The concentration of CO_2 is kept in equilibrium with sodium bicarbonate in the medium. Different media have differing buffering capacity. If a CO_2 controlled incubator is not available, or cultures must be kept sealed in flasks (*i.e.* after treatment with some volatile substances), then cells may be maintained in flasks sealed after gassing with 5% CO_2, 95% air, or vessels kept in boxes gassed and then sealed with pressure sensitive tape. In the case of boxes, the humidity must be maintained with a dish of water. Cultured cells can generally survive lower temperatures, but rarely survive temperatures greater than 2°C above normal, and therefore the incubator should be set to cut out at approximately 38.5°C to prevent cell death. Incubators are designed to regulate an even temperature and this is more important than accuracy, *i.e.* temperature should be ± 0.5°C. Most incubators have areas of differing temperature, therefore fan assisted incubators are preferable to help maintain even temperature distribution.

2.4.3 Inverted microscope

A simple inverted microscope is essential so that cultures can be examined in flasks and dishes. It is vital to be able to recognize morphological changes in cultures since these may be the first indication of deterioration of a culture. A very simple light microscope with 100x magnification will suffice for routine cell counts in a hemocytometer, although a microscope of much better quality will be required for chromosome analysis. A microscope with normal Kohler illumination will be needed for cell counting. An inverted microscope will also be needed for examining flasks and multi-well dishes from underneath. Microscopes should be equipped with a 10x and a 20x objective and it may be useful to provide a 40 x and a 100x objective for the normal microscope. Additional features such as a camera, CCD video camera, adapter and attachments and UV facility may also be required for some purposes.

2.4.4 Refrigerators and freezer (–20°C)

Both items are very important for storage of liquid media at 4°C and for enzymes (*e.g.* trypsin) and some media components (*e.g.* glutamine and serum) at –20°C. A refrigerator or walk-in cold room is required to store medium and buffers. A freezer will be needed for keeping pre-aliquot stocks of serum, nutrients and antibiotics. Reagents may be stored at a temperature of –20°C but if cells are to be preserved it may be necessary to provide a –70°C freezer or LN_2.

2.4.5 Liquid nitrogen/deep freezer

Invariably for continuous and finite cell lines, samples of cultures need to be frozen down for storage. It is important to maintain continuity in cells to prevent genetic drift and to guard against loss of the cell line through contamination and other disasters. The procedure for freezing cells is general for all cells in culture. They should be frozen in exponential phase of growth with a suitable preservative, usually dimethylsulfoxide (DMSO) or glycerol. The cells are frozen slowly at 1°C/min to –50°C and then kept at –196°C in liquid N_2. Deterioration of frozen cells has been observed at –70°C, therefore, –196°C (liquid N_2) seems to be necessary. To achieve slow freezing rates a programmable freezer can be used.

2.4.6 Water-purifying apparatus

The use of high purity, toxin-free water is essential for successful cell culture. Glass distillation or other water purification techniques (*e.g.* deionization) may be used but the water for cell culture should be autoclaved before use. One of the best sources of water for preparing cell culture media is pharmaceutical grade water for injection. All water for cell culture should be used as soon as possible after collection, as storage in any container will affect the purity of the water.

Double glass distillation

This was the first system to be used widely for cell culture and is still effective. A double distilled or reverse osmosis water supply is essential for preparation of media, and rinsing glassware. The pH of the double distilled water should be regularly checked as in some cases this can vary. Variations in the quality of water used may account for variation in results, therefore water from one source should be used. Water is sterilized by autoclaving at 121°C for 20 min. The distilled water must be glass distilled and stored in glass if it is to be used for the preparation of media. Storage in plastic may result in leaching of toxic substances from the plastic into the water. Distillation has the advantage that the water is heat-sterilized and of high purity but disadvantages include the large volume of water required for the distillation processes, slow production rate, high cost of operation and requirement for constant maintenance.

Deionization

Deionization requires a multistage process where the feed water is purified stepwise. The first step is usually reverse osmosis and subsequent steps can remove organic or inorganic materials (activated carbon) with further purifying steps (deionization) which may include membrane filtration depending on the quality of the feed water. The quality of the final product can be monitored using a conductance meter, which ideally should be around 18 megaohm. Advantages of the deioniser system are the high throughput of pure water, low wastage of water and the fact that the system can be modified according to the quality of the feed water.

2.4.7 Washing up and sterilizing facilities

The washing and sterilization facilities (autoclave, etc.) are often kept outside the tissue culture laboratory, since they would generate moisture (increased humidity) and heat.

Availability of a wide range of plastic tissue culture reduces the amount of necessary washing up. However, glassware such as pipettes should be soaked in a suitable detergent, then passed through a stringent washing procedure with thorough soaking in distilled water prior to drying and sterilizing. Pipettes are often plugged with nonabsorbent cotton wool before putting into containers for sterilization. Glassware, such as pipettes, conical flasks, beakers (covered with aluminum foil) are sterilized in a hot air oven at 160°C for one hour. All other equipments, such as automatic pipette tips and bottles (lids loosely attached) are autoclaved at 121°C for 20 min. Sterilizing indicators such as sterile test strip are necessary for each sterilizing batch to ensure that the machine is operating effectively. Autoclave bags are available for loose items. Aluminum foil also makes good packaging material.

2.4.8 Filter sterilization

The media may be purchased ready made or may be prepared in the laboratory, the latter being cheaper, if needed in large quantities. Media that cannot be autoclaved must be sterilized through a 0.22 μm pore size membrane filter. These are obtainable in various designs to allow a wide range of volumes to be filtered. They can be purchased as sterile, disposable filters, or they may be sterilized by autoclaving in suitable filter holders. Culture media, enzymes, hormones, cofactors and bicarbonate buffers are examples of non-autoclavable substances.

2.4.9 Facilities for counting cells

It is possible to monitor cell growth under microscope (looking for confluency), however, more accurate cell counts are required for most experimental purposes. The most commonly used device is the improved Neubauer hemocytometer originally designed for counting blood cells. An electronic particle counting method using a Coulter counter can also be used.

2.4.10 Balances, pH meter

A coarse and a fine balance and a simple pH meter are useful additions to the tissue culture area for the preparation of media and special reagents. A pH meter is required for preparation of stock solutions. Also an osmometer is essential to check solutions for osmolality.

2.4.11 Microscope

A good quality research microscope with regular bright field optics up to 100x objective magnification, phase contrast 40x/100x objective magnification, and fluorescence optics for fluorescent antibody observation is required in TC lab. Option of an automatic camera/CCD for photographic records should be available with microscope.

2.4.12 Temperature recording

Ovens, incubators, and hot rooms should be monitored regularly for uniformity and stability of temperature control. Recording thermometer which range for low temperature and also up to +200°C is required for monitoring frozen storage, cell freezing, incubators, and sterilizing ovens etc. Temperature and humidity display of TC lab is also required.

2.4.13 Roller racks

Roller racks of varying types are required for upscaling monolayer culture. The choice of apparatus is based on the size and number of bottles to be rolled at any given time based on the requirement of attainable cell density. The length of bottle range between 150 mm to 500 mm and also depend on size of the rack. The maximum surface is available when large number of bottles are used. The length of the bottle determine the maximum cell yield but is restricted by the size of the rack and the height of the rack will further determine the number of rows of bottles used.

2.4.14 Simple pipetting

Simple pipetting is one of the most frequent tasks required in the normal handling of routine culture. While a small-scale satisfactorily handling is achieved with a rubber bulb but speed, accuracy, and reproducibility are greatly enhanced by using a pipetting aid. These pipette aids have a filter placed at the pipette insert to avoid contamination. Disposible sterile pipettes are used with ease without risk of contamination.

2.4.15 Micropipettes

Micropipettes are used extensively for dispensing small volume of liquid. The tip used is sterile, thus if a sterile fluid is withdrawn from a container with a micropipette, the non-sterile part of the pipette must not touch the sides of the container. Small volumes of liquids can be handled accurately with micropipettes.

2.4.16 Culture vessels

The actual choice of a culture vessel depends on factors *viz.* (i) culture in monolayer or suspension; (ii) quantity of cells required and purpose; (iii) frequency of sampling for the required work and (iv) cost factor and economy. Cell yield is directly proportional to available surface areas. Anchorage dependent cells require a nontoxic, biologically inert, and optically transparent surface that will allow cells to attach and grow. The most convenient vessels are specially treated polystyrene plastic that are supplied sterile and are disposable. These include petri dishes, multi-well plates, micro titer plates, roller bottles, and screw cap flasks T-25, T-75, T-150 (cm^2). Tissue culture ware is always chosen to match the procedure. Sometimes it may be necessary to condition a surface by pretreatment with 'spent' medium which has been used with another culture (conditioned medium). Although all tissue culture plastic ware should support cell growth adequately, but tests should be conducted for new source or type of dish to ensure that cultures grow readily in it.

2.4.17 Colony counter

A variety of colony counter from electrode tipped marker pen, programmable electronic counter or automatic counters are available for monolayer colony counting.

2.4.18 Controlled-rate freezer

Some cells may be frozen by simply placing them at –70ºC, while other cells may require different cooling rates. A programmable freezer from reputed manufacturer enables the cooling rate to be controlled by preset program.

2.4.19 Flow cytometer

This instrument, also known as a fluorescence activated cell sorter (impulse cytophotometer or cytofluorimeter), can analyze cell populations and separate them according to a variety of criteria. It has almost unlimited potential but is too expensive to come within most tissue culture laboratory equipment budgets. Its major advantages include sophisticated cell population analyses and isolation of minority constituents of a cell population by single or multiple criteria.

2.5 Laboratory Operation

The important principles and approaches that may be adopted to ensure satisfactory operation of a cell culture laboratory includes the following important points:

- Only essential personnel should have access to cell culture areas.
- Cell culture areas should be dedicated and separate laboratories designated for other work.
- Laboratory layouts should allow for easy movement of personnel working in the TC lab. Each cell culture work station should be organized such that all items needed are available readily to hand while handling cells.
- The use of sinks in the cell culture area should be avoided since these can be a major source of microbial contamination.
- Liquid nitrogen storage areas should be well ventilated.
- Standard operating procedures should be in place and functional for all procedures/operations in the tissue culture lab.

CHAPTER 3

STERILIZATION PROCEDURES

Culture procedures involve preparation and sterilization of glassware, equipment, reagents and media. They also involve preparation of animal tissue that needs to be cultured. The apparatus and reagents meant for tissue culture should be reserved for that purpose alone. Washing of glassware should be separate, as it requires different and more elaborate washing steps.

3.1 Sterilization

Sterilization and decontamination

Sterilization can be defined as any process that effectively kills or eliminates transmissible agents (such as fungi, bacteria, viruses, spores and prions) from a surface, equipment or biological culture medium. Getting rid of contamination requires sterilization. Sterilization may be achieved by: (i) Physical destruction of contaminants (dry heat, autoclaving, irradiation), (ii) Physical removal (*e.g.* filtration of solutions such as medium) or (iii) Chemical destruction (*e.g.* antibiotics). The success of the sterilization process depends upon the choice of the method adopted for sterilization.

Several factors those affect sterilization are:

1. The dryness of articles/devices to be processed
2. The temperature and humidity of the processing area
3. Proper preparation and loading of articles into the sterilizer
4. Delivery of sterilizing agent into the system
5. Sterilizer's condition and maintenance protocol
6. Use of correct sterilization method and cycles followed

3.2 Methods of Sterilization

Table 3.1 Sterilization Methods used for Different Materials

	Methods				
Material	***Dry heat***	***Autoclave***	***Radiation***	***Gas***	***Filtration***
Metal	✓	✓	✓	–	–
Glass	✓	✓	✓	–	–
Plastics					
Polycarbonate	–	✓	✓	✓	–
Polyethylene	–	✓	✓	✓	–
Polypropylene	–	✓	✓	✓	–
Polystyrene	–	–	✓	✓	–
Media	–	–	✓	–	✓
Serum	–	–	✓	–	✓
Salt solutions	–	✓	✓	–	✓

3.2.1 Dry heat sterilization

The principle of dry heat sterilization is conduction of heat from the exterior to the interior of an object in the absence of moisture. It employs higher temperatures in the range of 160–180°C and requires exposures time up to 2 hr, depending upon the temperature employed. This form of sterilization is ideal for oily substances, powders, metals, and most glassware while it is unsuitable for fabrics or rubber articles. The benefit of dry heat includes good penetrability and non-corrosive nature which makes it applicable for sterilizing glasswares and metal surgical instruments. Dry heat destroys bacterial endotoxins (or pyrogens) which are difficult to eliminate by other means and this property makes it applicable for sterilizing glass bottles which are to be filled aseptically. The main disadvantage of dry heat is that it is not a rapid means of sterilization. The oven should not be overloaded and the load should be standardized to ensure equal heating of the load.

Hot air ovens

Hot air ovens are used for sterilizing glassware, oils, dry powders, etc. A variety of ovens are available for use, however, an oven with forced air circulation is preferred. A temperature range of 160–180°C maintained for one to two hours is generally acceptable for sterilization. It must be remembered that the center of the load is least accessible and space must be left between items. Small ovens may prove to be advantageous, as they allow for more rapid and uniform heating apart from being economical when a small load is required.

The advantages of dry heat sterilization are as follows:

- Can be used for almost any type of glassware
- Eliminates the possibility of steam related contamination

- Means of sterilizing items that cannot be sterilized any other way, *e.g.* powders and oils.

The disadvantages are:

- Time consuming
- Unsuitable for certain items, such as fabrics and rubber
- Can be difficult to control because of the presence of cool spots in oven and some items are poor conductors of heat.

Preparation of materials

In preparing glasswares *e.g.* pipettes for dry heat sterilization, the first step is to plug the back ends with cotton batting. Pasteur pipettes are put into their glass containers and graduated pipettes are sorted and put into their appropriately labeled metal canisters. Glass petri-dishes may be wrapped in foil or paper designed specifically for dry heat sterilization. Test-tubes have their orifices covered in foil and are inserted into test tube racks. Alternatively, they may be sterilized, without individual wrapping in the foil. Beakers and graduated cylinders likewise have their open ends covered with foil during sterilization and storage. Glass cover slips are sterilized keeping in glass petri-dishes.

Alcohol and flaming

Swabbing with alcohol (ethanol) and flaming are techniques commonly used for sterilization in tissue culture procedures. Alcohol is a known surface disinfectant. Swabbing work surfaces with 70% alcohol helps in removing dust and keeping the area clean.

When wiping a work surface or various other items with alcohol, simply wipe from one side to another working from the top to the bottom of the surface. Similarly, a bottle is wiped in a spiral fashion from the neck area toward the base. During tissue culture procedures, any spillage should be wiped up by a tissue/cotton dampened with alcohol. This keeps the area clean and prevents aerosols.

Ninety five percent alcohols in conjunction with flaming is an effective means of sterilization. The necks of vessels are flamed all the way around before and after removal of their caps. Culture vessels such as T-flasks and other flasks are flamed before and after loosening the cap. Similarly, beakers, flasks and glass measuring cylinders are flamed around their orifices before use. The neck of a bottle or vessel is flamed once more before replacing the cap.

3.2.2 Steam sterilization

Of all the methods available for sterilization, moist heat or steam under pressure is the most widely used and the most dependable. Steam under pressure is used to generate high temperature needed for sterilization. Steam sterilization is nontoxic, inexpensive, rapidly microbicidal, sporicidal, and penetrates well, giving good results. The basic principle of steam sterilization is to expose each item to direct steam contact at the required temperature and pressure for the specified time. Saturated steam (steam in thermal equilibrium with water) acts as an effective sterilizing agent. Pressure serves as a means to obtain the high temperature necessary to quickly kill microorganisms. Specific temperatures must be obtained to ensure the microbicidal activity. The most common steam sterilizing temperatures is 121°C while the range is 121°C to 134°C and the time usually between 15 to 30 minutes at 15 pound pressure.

Autoclave

Autoclaves use pressurized steam to destroy microorganisms, and are the most dependable systems available for the sterilization of laboratory glassware, media, and reagents. Material to be sterilized is placed into a chamber that can be filled with steam under pressure. Usually the autoclave is set to give 121°C which can be achieved at 5 kg/cm^2 (15 lbs/in^2) of steam. Exposure to these conditions for 15–20 minutes will sterilize most materials, but larger volumes and bulk will require longer period. The method is effective because of the high temperatures that can be reached and because moist heat has greater penetrating power than dry heat. It is ideal for liquids as steam under pressure prevents boiling and evaporation.

Advantages are: 1) Rapid sterilization, 2) May be used for sterilizing rubber and fabric, 3) Does not leave a toxic residue provided water used for production of steam is of good quality.

Preparation of materials

The mouth of glass wares are covered with foil and then secured with brown paper and thread. Tip boxes are covered with foil and paper. Small items may be placed into petri-dishes then wrapped in foil. Bottles with screw caps are autoclaved with the cap slack and covered with foil. The caps are tightened through the foil when the bottles have cooled after sterilization.

Sterilization indicators

Sterilization indicators are a useful adjunct to autoclaving and dry heat sterilization. Paper sterilization indicators are chemically treated such that they change color when exposed to the correct combination of temperature with moisture and time. In general, autoclave tape ensures which articles have been autoclaved and which have not. Paper indicators are considered to be good indicators of sterility as they provide the results far more rapidly than biological tests.

3.2.3 Physical removal: filtration

There are a variety of filters available, most designed for specific purposes. The choice of which type of filter to be used depends on following considerations:

- Pore size of filter
- Chemical compatibility
- Hydrophobic *vs* hydrophilic filter
- Toxicity of filter

A pore size of 0.2 μm provides the sterilization necessary for tissue culture media. Chemical compatibility refers to the compatibility of what is to be filtered. For sterilizing media or other aqueous solutions, a hydrophilic filter is required. Hydrophobic filters are ideal for sterilizing gases, for example CO_2.

Membrane filtration

For heat sensitive liquids like vitamins and other reagents and solutions, membrane filters are generally used and are composed of synthetic material, *e.g.* cellulose acetate, polycarbonate, etc., that have been constructed with pores of specific size (0.22–0.45 μm). Millipore filters (a brand name) are commonly used for sterilization of liquids through filtration.

3.2.4 Radiation

Many types of radiation are used for sterilization like electromagnetic radiation (*e.g.* gamma rays and UV light), particulate radiation (*e.g.* accelerated electrons). Gamma rays and electrons cause ionization and free radical production while UV light causes excitation. (250–270 nm) helps to keep work areas sterile when not in use. The optimum wavelength for UV sterilization is 260 nm. A mercury lamp giving peak emission at 254 nm is the suitable source of UV light in this region. A 2 hr exposure to UV light is usually sufficient to kill microorganisms. Because UV light will not penetrate the surface of plastic or glassware, it can only be used with open dishes that are directly exposed to the UV source.

γ-Radiation

Gamma radiation is widely used commercially to sterilize plastic ware and lyophilized reagents. It is not commonly used in culture laboratories, because it requires a γ-radiation source generator and appropriate shielding of the radiation.

3.2.5 Gas sterilization

Temperature sensitive materials, such as plastic ware, can be sterilized by exposure to toxic gas under high pressure. Ethylene oxide is commonly used to sterilize plastic ware. Because the principle of this method of sterilization is that the toxic gas is forced into the pores of materials, items sterilized by this method can not be used until the toxic gas has re-equilibrated and is out of the materials (out gassing). The length of time for out gassing depends on the apparatus ranging from 1–2 days to 1 week. Minute amounts of toxic gas may be harmful to the cultures, therefore, it is safer to wait the recommended time before using the sterilized equipment.

3.2.6 Chemical sterilization

Different detergent are to be tested for their (i) ability to wash heavily soiled glassware, (ii) quality of the growth surface after washing, and (iii) toxicity of the detergent.

Chromic acid

Chromic acid, is a hazardous substance, with safety and environmental concerns. There are effective commercially available substitutes to chromic acid. Some heavily soiled glassware may require use of chromic acid (10% potassium dichromate in 25% sulphuric acid). If chromic acid is to be used, follow all normal safety precautions for using concentrated acids and acid solutions. As with any other cleaning process, all cleaning solutions must be completely rinsed from the glassware through copious changes of tap water followed by several changes of distilled water.

Caustic alkaline detergents

Caustic alkaline detergents render the surface of the glass unsuitable for cell attachment and require subsequent neutralization with hydrochloric acid and sulphuric acid.

Sodium hypochorite

This is a good sterilizing agent which helps 1) to remove any potential biohazard, and 2) to prevent microbial contamination growing up in the water.

3.3 Sterilization of Tissue Culture Glasswares

Plastic wares are supplied sterilized and ready for use which are disposable in nature and are not reused. They are costlier than glasswares but are adequate for most cell culture purposes. However, glasswares are reused, and hence most economical (Table 3.2) but must be carefully washed following standardized procedures.

- Rinse all glassware as soon as possible after use.
- Store soiled items in water containing a disinfectant or cleanser to avoid drying and making items harder to clean.
- Use a recommended detergent for cell culture glass ware washing.
- Clean glass by scrubbing with a brush. Periodically inspect brushes for wear to avoid scratching glass.
- Thoroughly rinse items in tap water, followed by at least 7 changes of distilled water and 4 changes of deionized water. Even the smallest residual amounts of cleansers, disinfectants or acids can affect the growth of cells in cultures.
- Dry glassware on racks or peg boards and inspect after drying. If glassware is hazy, has a film or patches, then additional cleaning is required before use.

Sterilize cell culture glassware using a hot air oven at 160°C for three hours to destroy pyrogens. Non-glass components which may not withstand 160°C should be sterilized by alternate methods such as autoclaving.

Table 3.2 Sterilization Methods of some Commonly used Articles in Cell Culture

Item/article	*Sterilization method*
Tips for micropipettes	Autoclave
Reusable filter assembly	Autoclave
Tissue culture glasswares	Dry heat
Glass bottles with screw caps	Autoclave
Glass slides and coverslips	Dry heat
Glass syringes	Autoclave
Magnetic stirrer bars	Autoclave
Glass pipettes	Dry heat
Glass Pasteur pipettes	Dry heat
Screw caps, reusable	Autoclave
Silicone tubing	Autoclave
Glass test tubes	Dry heat

Autoclave = 100 kPa (15lb/cm^2); 121°C for 20 min
Dry heat = 160°C/1 hr

Screw caps/stoppers

Two types of screw caps for glass bottles are in common use: (1) aluminum or phenolic plastic caps with synthetic rubber or silicone liners and (2) reusable polypropylene caps, which are deeply shrouded and have ring inserts to improve the seal. Use silicone or heavy metal-free white rubber stoppers in preference to natural rubber.

Place caps in a glass petri-dish with the open side down. Autoclave for 20 min at 121°C.

Sterilizing filters

Reusable filter holders should be made up and sterilized as follows:

1. After thorough washing in detergent, 7 times rinse in distlled water, then 4 times in deionized water, and dry.
2. Insert support grid in filter and place filter membrane on grid.
3. Place pre-filters on top of filter, reassemble filter holder, but do not tighten up completely.
5. Cover inlet and outlet of filter with aluminum foil.
6. Pack filter assembly in sterilizing paper and place autoclave indicator tape.
7. Autoclave at 121°C, 15lb/in^2, remove and allow to cool.

Miscellaneous equipments

Cleaning

All new apparatus and materials (silicone tubing, filter holders, instruments, etc.) should be soaked in detergent overnight, thoroughly rinsed, and dried. Anything that will corrode in the detergent *viz.* mild steel, aluminum, copper, or brass, etc., should be washed directly by hand, without soaking, or soaking only for 30 min. Used items should be rinsed in tap water and immersed in detergent immediately after use.

Packaging

Ideally, all apparatus for sterilization should be wrapped in a covering that will allow steam penetration but be impermeable to dust, micro-organisms etc. Proprietary bags are available bearing sterile indicating marks that show up after sterilization. Tubes and orifices should be covered with tape and paper or nylon film before packaging, and needles or other sharp points should be shrouded with a glass test tube or other appropriate guard.

Sterilization

The type of sterilization used will depend on the material (Table 3.1). Metallic items are best sterilized by dry heat. Silicone rubber, teflon, polycarbonate, cellulose acetate, and cellulose nitrate filters, etc., should be autoclaved for 20 min at 121°C, 100 kPa (1 bar, 15 Ib/in^2). In small bench-top autoclaves and pressure cookers, make sure that the autoclave boils vigorously for 10–15 min before pressure to displace all the air. Enough water is put in at the start to allow for this. After sterilization, the steam is released and the items removed to dry off in an oven or rack. Take care releasing steam and handling hot items to avoid burns. Wear elbow-length insulated gloves and keep face well clear of escaping steam when opening doors, lids, etc. Use safety locks on autoclaves.

3.4 Sterilization of Tissue Culture Reagents

Table 3.3 Sterilization of Commonly used Liquids/Reagents in Cell Culture

Solution/reagents	*Sterilization method*
Amino acids	Filter sterilization
Antibodies	Filter sterilization
Bovine serum albumin	Filter sterilization
EDTA	Autoclave
Glutamine	Filter sterilization
Glycerol	Autoclave
HEPES	Autoclave
$NaHCO_3$	Filter sterilization
NaOH	Filter sterilization
Phenol red	Autoclave
Salt solutions (without glucose)	Autoclave
Serum	Filter sterilization
Trypsin	Filter sterilization
Vitamin stocks	Filter sterilization
L- Glutamine	Filter sterilization
Water	Autoclave

Reagents

Most reagents or media can be sterilized either by membrane filtration (if heat labile) or by autoclaving if they are heat stable (Table 3.3).

Water

Water must be of a very high purity for use in tissue culture. As water supplies vary greatly, the degree of purification required may vary. There are four main approaches to water purification: distillation, deionization, carbon filtration, and ultra filtration. Simple systems depend on the first two alone or combined, while the more efficient systems employ multiple stages, each operating on a different principle.

Serum

Serum is usually sterilized by filtration through a 0.1 μm-porosity sterilizing filter, but because of its viscosity and high particulate content, it should be passed through a graded series of cartridge type glass-fiber prefilters before passing through the final sterilizing filter. The prefilters assemblies may be stainless steel with replaceable cartridges or may be a single bonded unit. For smaller volumes (less than 1 liter) graded filters may be stacked in a single unit.

3.5 Sterilization of Media

Concentrated media are often prepared at a low pH (between 3.5 and 5.0) to avoid precipitation and keep all the constituents in solution. The preparation of media is rather complex. It is convenient to make up a number of concentrated stocks. The requisite amount of each concentrate is then mixed and finally filtered through 0.2 μm porosity filters.

Preparation and sterilization of medium

The various media constituents and other reagents used in cell cultures must be carefully sterilized (Table 3.3) either by autoclaving or by filtration. Heat stable constituents like water, salts, supplements like peptone or tryptose etc. are autoclaved at 121°C for 20 min. But heat labile constituents like serum, trypsin, proteins, growth factors etc. must be sterilized by filtration through a optimum (0.2–0.5μm) porosity membrane filter. Each filtrate should be tested for sterility to avoid failure due to contamination.

An autoclave cycle that ensures a sterilizing temperature of 121°C for at least 15 minutes at a pressure of 15 psi (100 kPa) is adequate for liquid reagents. Some reagents, such as glucose solutions, may have some degree of heat resistance and may be sterilized in an autoclave at 115°C, 10 psi (70 kPa) for 15 minutes.

Filtration of readymade media

Powdered media or salt mixtures are extremely hygroscopic and should be protected from atmospheric moisture. The entire contents of each package should be used immediately after opening. Preparing media in concentrated form is not recommended. Some free base amino acids have low solubility coefficients and insoluble salt complexes may precipitate in concentrated solution. Supplements can be added prior to filtration or introduced aseptically to sterile media. The nature of the supplement will affect the storage conditions and shelf life of the medium.

3.6 Quality Control of Sterilization

Routine labeling and record keeping

The importance of keeping track of reagents for cell culture will help to determine whether it is a sporadic contamination induced while working or it is part of a failure in sterilization procedure. Maintaining "ideal" records would indicate how and when all glassware and equipment were sterilized, when and how any solutions were prepared and sterilized. If the same type of contamination occurs, these records are the tools to help and determine whether any glassware, equipment, or reagent were common to both occurrences of contamination.

Routine sample analyses (test cultures)

It is desirable to routinely test solutions for sterility before adding them to cultures. To do so, remove an aliquot of the solution and incubate the material at 37°C and at room temperature in an area that is isolated from cultures. The aim of this is to identify any microbial growth in the test samples before they are used as working solutions. If contamination is present, the solutions will become turbid, perhaps as rapidly as 2–3 days after inoculation.

Equipment checks

Proper equipment function is essential to sterility. Keep a direct access on oven or autoclave, periodically check the outside gauges to ensure that the temperature and pressure are correct. If there are no oven gauges use a thermometer and check the temperature at some time during the sterilization. In addition, each sterilization run should contain an indicator on some or all the materials. Both thermal indicators for ovens and autoclave tape are commercially available which should be regularly used for effective monitoring.

CHAPTER 4

CULTURE MEDIA AND REAGENTS

4.1 Introduction

Animal cells *in vivo* are provided with nutrients and growth factors by the vascular system. The extra cellular environment helps to maintain the viability of the cells and their growth functions. There have been constant efforts to develop culture systems as close to natural vascular system to grow cells *in vitro* system since the first time in 1885 when chicken cells were grown outside the body system. To grow cells *in vitro* a similar control of the extra cellular environment is necessary. Thus, culture systems are designed to mimic the *in vivo* environment in laboratory. Culture medium is the most important pivot for culturing cells and tissues. Animal cells need either a completely natural medium or an artificial medium supplemented with all the essential nutrients required for growth and other functions which they are mandated.

Since 1950s, research in tissue culture media has lead to development of conditions, which closely simulate the situation *in vivo*. Today many decades of research have lead to standardized media and sophisticated incubation conditions, for culturing animal cells *in vitro*. The environment is regulated with regard to the temperature, osmotic pressure, pH, essential metabolites (such as carbohydrates, amino acids, vitamins, proteins and peptides), inorganic ions, hormones and extracellular matrix. Among the biological fluids that proved successful for culturing cells, serum is the most significant; 5 to 20% of serum is usually added to media for optimal cell growth. Serum is an extremely complicated mixture of compounds including many undefined components.

4.1.1 Properties of media and salt solutions

Media are designed in such a way that it provides the optimum conditions (pH, osmotic pressure, chemical constituents), for cells or tissues that are required for their growth and maintenance in culture *in vitro*.

The substrate that maintains cells for long periods of time (days to weeks) is referred to as culture medium. The cell membrane is semi permeable, thus, water and some of the intracellular molecules can freely pass through the membrane. Examples of common isotonic solutions are buffered salines (Dulbecco's) and balanced salt solutions (Hank's, Earl's).

If the outside fluid has a greater concentration of solute molecules than the cell (2 M NaCl solution), it is hypertonic and will cause the water inside the cell to leave. This causes shrinkage of the cells and may induce irreparable damage to the cells. On the contrary, a solution that has a lower concentration of molecules than the cell (10 mM NaCl solution), will cause water to enter the cell. The cell will swell and eventually burst. Normal cell survival requires that the fluids have a pH that is compatible with the cell. In general, this pH value is 7.4, although there are cell-specific variations, like pH values closer to 7.0 are better for the culture of human peripheral blood monocytes and Chinese hamster ovary (CHO) cells. Maintaining the correct pH is accomplished by the addition of a buffer. As carbon dioxide is normally produced by cells, bicarbonate buffer is the most commonly used buffer. Non-bicarbonate buffers such as HEPES have been used successfully (with or without bicarbonate) in many cell systems. Other organic buffers such as MOPS have been included in media formulations. The advantage of the organic buffers is that they allow a stricter control of the pH in the system while they may sometimes be toxic to cells.

4.2 Types of Media

The nutrient media used for culture of animal cells must be able to support their survival as well as growth. The various types of media used for tissue culture may be grouped into two broad categories: (i) Natural media and, (ii) Artificial media.

The choice of cell culture medium is extremely important, and significantly affects the success of cell culture experiments. The choice of medium depends mainly on the type of cells to be cultured (normal, immortal or transformed), and the objective of culture (growth, survival, differentiation or production of desired proteins). Different cell types have highly specific growth requirements, and the most suitable medium for each cell type must be determined experimentally. Common basal media include Eagle minimal essential medium (MEM), Dulbecco's Modified Eagle Medium (DMEM), RPMI 1640, and Ham F10. These medium contain a mixture of amino acids, glucose, salts, vitamins, and other nutrients, and are available either as a powder or as a liquid from various commercial suppliers. Non transformed or normal cells (finite life span) and primary cultures from healthy tissues require defined quantities of proteins, growth factors and hormones. In contrast, transformed cells synthesize their own growth factors, but may require factors like insulin, transferrin, lipids etc.

Basal media are usually supplemented just before use with serum, L-glutamine, and antibiotics and/or fungicides to give complete medium (also called growth medium). Serum contains growth and attachment factors, and may show considerable variation in the ability to support growth of particular cells. Fetal calf serum (FCS) is the most frequently used serum in cell culture experiments. Different serum batches should be tested to find the best one for each cell type. L-glutamine is an unstable amino acid that, with time, converts to a form that cannot be used by cells, and should be added to medium just before use. Antibiotics and fungicides can be used as a supplement to aseptic technique to prevent microbial contamination. Some cell types, particularly primary cells, require additional supplements (*e.g.* collagen and fibronectin, hormones and growth factors) to attach to the cell culture vessel and proliferate.

4.2.1 Natural media

Natural media is the most preferred medium, since it is the cheapest and most convenient. The natural media, which promote cell growth, fall into the following three categories.

Blood plasma

After many years of research it was found that plasma provided a complete nutrient in which cells could survive and multiply slowly for extended periods under conditions that resembled *in vivo* conditions. Plasma is still being used for the following purposes:

1. To provide a nutritive substrate and a supporting structure for many types of cultures, just as it also provides a matrix for new cells during the repair of injury in the body.
2. Conditioning the surface of glass for better attachment of cells.
3. Protecting cells and tissues from excessive traumatic damage during sub culture.
4. Protection from sudden changes in the environment at times of fluid change.
5. Formation of localized pockets of conditioned medium around cells.

For culture work, plasma from the adult chicken is preferred to mammalian plasma because it forms a clear, solid coagulum even when diluted several times. Plasma clots are now available commercially either as liquid plasma in siliconized ampoules or as lyophilized plasma.

Biological fluids such as serum

Blood serum (plasma minus fibrinogen) with or without other nutritive substances may be used either as the entire culture medium or as the fluid phase of a medium consisting partly of a plasma coagulum. For many years it was assumed that whole serum was toxic, and plasma was useful only as a supportive structure and that the nutritive requirements of the cells were supplied by the embryo extract that was usually added to the medium. Eventually, however, it was found possible to cultivate tissues in serum alone without plasma or tissue extract. As some of the more elaborate chemically defined solutions were developed, it was found that they had to be supplemented with 10 to 20% serum to provide completely adequate medium for the continuous propagation of established cell lines and freshly explanted tissues for extended periods. Harris (1959) concluded that medium 199 and NCTC-109 and the simpler basal medium of Eagle (1955) are all deficient in one or more factors that occur in serum dialysate and are essential for the growth and maintenance of chick skeletal muscle fibroblasts. Thus serum does provide some of the growth factors or some of the physical conditions, or both, that are presently lacking in synthetic media.

Serum is obtained from human (adult blood, placental cord blood) horse blood or calf blood. Out of these, placental cord serum and foetal calf serum are satisfactory. The serum can be obtained as exuded liquid from blood undergoing coagulation and is filtered using millipore filters. The serum should be tested for sterility and toxicity before using and stored at low temperature. The presence of contaminants can be checked by incorporating the serum in a growth medium. The toxicity of sera can be reduced by heat inactivation.

4.2.2 Artificial media

The various artificial media developed for cell culture may be grouped into four classes:

A. Serum containing media

The essential amino acids histidine, isoleucine, leucine, lysine, methionine, phenylalanine, threonine, tryptophan, valine, cysteine, glutamine, and tyrosine (all at 10^{-4} to 10^{-5} M).

***Vitamins*:** Choline, folic acid, nicotinamide, pantothenate, pyridoxal, thiamine inositol and riboflavin.

Salts: Na^+, K^+, Ca^{2+}, Mg^{2+}, Cl^-, PO_4^{3-}, HCO_3^-

Glucose: 0.9 g/L

Dialyzed serum: 5–10% of total volume.

B. Serum free media

Ever since the observation in the 1950s that natural media could be replaced in part by synthetic media, attempts have been made to culture cells without serum. Scientist further modified their formulation by adding insulin and with other minor modifications, were able to culture HeLa cells without serum. Specific formulations (*e.g.* MCDB 110) have been derived in the absence of serum to culture human fibroblasts (Ham, 1984), many normal and neoplastic murine and human cells, lymphoblasts and several different primary cultures. While in some cases, some protein is added, on the other hand for many individual cell types, the use of serum-free media is difficult and not always straightforward.

Taking together all reasons and benefits for the use of serum-free media, there are certainly advantages of this approach for those cell lines, which are designed for the production of cell proteins or especially in the stem cell research.

C. Chemically defined media

Earlier, the nutritive media for the cultivation of animal cells *in vitro* were consisted of blood plasma, blood serum, tissue extracts, etc. The complexity and variability of these naturally occurring materials made it difficult to use them in experiments designed to determine the nutritive substances required by the cells and the effect of particular substance upon them. The first attempts to devise chemically defined media were made by Lewis and Lewis (1911), Baker and Carrel (1926). Carrel's discovery (1912) that extracts of embryonic tissues contained an abundance of growth promoting substances made it possible for the first time to propagate animal cells indefinitely and stimulated the search for the particular substances responsible for the same. Baker and Carrel (1926) reported that Witt's peptone, proteases and other degradation products when used with serum or plasma provided essential nutrients for fibroblasts, epithelial cells and blood monocytes.

White (1946) for the first time made serious attempt to cultivate animal tissues in a complete solution of known composition. He reported a feeding solution of 20 ingredients of known composition that supported relatively large masses of chick embryo heart tissue in a state of functional survival for several weeks. Morgan *et al.* (1950) published the composition of a more adequate medium number 199. The medium included almost complete complement of amino acids and vitamins as well as several nucleic acid constituents and certain intermediary metabolites and accessory growth factors. Fischer's group included glutamine because of its importance place in the medium.

These studies allowed the development of media formulations to replace the biological extracts previously used. Defined media had the advantages of consistency between batches of media, ease of sterilization and a reduced chance of adverse microbial contamination. Media formulations used today include a complex mixture of carbohydrates, amino acids, salts, vitamins, hormones and growth factors. These media contain contamination-free ultra-pure inorganic and organic constituents, and may contain pure protein additives, like insulin, epidermal growth factor etc. that have been produced in bacteria or yeast by genetic engineering.

D. Defined media

Although natural media are very useful and convenient for a wide range of uses, they suffer from disadvantage of poor reproducibility due to lack of knowledge of the exact constitution. For this reason, synthetic media have been designed for a variety of tissues and purposes, (i) media for immediate survival, (ii) media for prolonged survival, (iii) media for indefinite growth, and (iv) media for specialize functions. For immediate survival, a 'balanced salt solution' with defined osmotic pressure and pH is needed. This requirement is met by combination of certain inorganic ions, salts and glucose. For longer survival, serum may be used or the balanced salt solution may be supplemented with amino acids, oxygen, vitamins, and serum protein. One such medium was developed and modified by Eagle (1955), which is described as minimum essential medium (MEM). This contains essential amino acids, vitamins and salts. Other more complex synthetic media (i) 199 (Morgan *et al.*, 1950), (ii) CMRL 1066 (Parker *et al.*, 1957), (iii) RPMI 1640 (Moore *et al.*, 1967), (iv) F12 (Ham, 1965), (v) McCoy's 5 A, and (vi) Iscove's modified Dulbecco's medium (IMDM). All these defined media are usually supplemented with 5–20% serum. Often a 1:1 mixture of DMEM and F12 is used as a serum-free formulation.

E. Protein free media

Protein-free media do not contain any protein; they only contain non-protein constituents necessary for culture of the cells.

F. Selective media

One of the major advantages over growth-promoting activity afforded by serum-free media is in the ability to make a medium selective for a particular cell type. Separate lineages and even stages of development may be selected in hemopoietic cells by selecting the correct growth factor or group of growth factors. In addition to this the possibility of switching from a growth factor, after necessary amplification of the culture, to a differentiation factors is possible. The amplified culture may then be made to perform one or more specialized functions (Table 4.1).

Table 4.1 Growth Factors Promoting the Proliferation of Specific Mammalian Cells in Culture

Factor	*MW*	*Comment*
Epidermal growth Factor (EGF)	6400	Enhances DNA synthesis in quiescent ectoderm and endoderm derived cells
Fibroblast growth Factor (FGF)	13400	Composed of a single polypeptide chain which stimulates mesoderm cells
Platelet-derived Growth factor (PDGF)	30000	Principal mitogen in mammalian serum for cells of mesenchyme origin
Interleukin-l (IL-l)	17000	Stimulates interleukin-2 Production by T cells
Interleukin-l (IL-2)	17000	Stimulates the propagation of Human Lymphocytes
Nerve growth Factor (NGF)	130000	Promotes the secretary activity of neurons
Erythropoietin	23000	Stimulates erythroid cells and may be used in the treatment of anemia

4.3 Selection of Media

Different cell types/lines requires different media. Various continuous cell lines (HeLa, BHK21), primary cultures of human, rodent, and chicken fibroblasts, can be maintained on a relatively simple medium such as Eagle's MEM, supplemented with calf serum. More complex media are required for expression of specialized function, when cells are passaged at low seeding density 10^3/ml, as for cloning. These more demanding culture conditions that require complex media also require fetal bovine serum than calf or horse serum. Selection of the appropriate medium for a given cell type therefore depends on function (Table 4.2).

Several assays have been developed for the determination of the best medium for a given cell type: (i) long-term (several days) cell multiplication assay in form of clonal growth assay, cell growth curve analysis need to be done, and (ii) short term (several hr) [^{3}H]-thymidine assay is utilized. It is desirable to use the long term assays.

Preparing culture media

Most cell culturists do not prepare media from individual components but this can be done by same accuracy and care (Table 4.3). All reagents should be of the highest quality or be approved for cell culture use by the manufacturer. The water used must be pure. After components are added the pH should be adjusted to the required amount and the solution brought to its final volume and filtered immediately. Many media formulations include phenol red which serves as a color indicator of the pH of the solution. It is non-toxic to most cells and allows to estimate the pH of a solution or culture by simply looking at it. At pH levels below pH 7.0, solutions containing phenol-red appear orange; at pH 7.2–7.4 they are red; and above pH

7.5 they are reddish-blue, bordering on purple. Keeping in mind the use and the shelf life, quantity of media to be prepared is fixed.

Table 4.2 Choosing the Appropriate Medium and Culture System

Medium	*Applicable to*
Basal medium eagle(BME)	Growing cells with Serum
Minimal essential medium (MEM)	Growing cells with dialyzed serum
Dulbecco's Modified Eagles Medium (DMEM)	Many Virus transected cells, growth with serum high density growth
Ham's P-10 medium	Chick embryo cells, serum
Ham's P-12 nutrient mixture (F-12)	Chinese hamster ovary cells, low density, low-serum protein
William's medium	Rat liver epithelial cells
RPMI - 1630	Mouse leukemia cells, cells in suspension
RPMI - 1640	Human leukemia (and other) cells, hybridoma
Fischer's medium	Murine leukemia cells
McCoy's 5A medium	Human lymphocytes
MCDB 131	Human endothelial cells
Medium 199	Chick embryo fibroblasts
Medium NCTC 109	Hybridoma
Medium NCTC 135	Serum, serum-free growth
Neurobasal medium	Central nervous system neurons

Table 4.3 Ingredients of Media

Amino acids	*mM*	*Vitamins*	*uM*	*Salts*	*mM*	*Other materials*	
Arqinine	0.1	Biotin	1	NaCl	100	Glucose	5 mM
Cysteine	0.05	Choline	1	KCl	5	Penicillinin	0.005%
Glutamine	2.0	Folic acid	1	NaHO4	1	Streptomycin or	0.005%
Histidine	0.05	Nicotinamide	1	HCO 3	20	Gentamycin	0.005%
Isoleucine	0.2	Pantothenic acid	1	CaC12	1		
Leucine	0.2	Pyridoial	1	MgC12	0.5		
Lysine	0.2	Thiamine	1				
Methionine	0.05	Riboflavin	0.1				
Phenylalanine	0.1						
Threonine	0.2						
Tryptophan	0.02						
Tyrosine	0.1						
Valine	0.2						

4.3.1 Components of media

(a) Basic media

The most basic media are balanced salt solutions (BSS), viz. phosphate-buffered saline (PBS), which may be used for washing cells and for short incubations in suspension. More complex defined media are used for growth and maintenance. Defined media can also vary in complexity, by the addition of a number of constituents, *e.g.* from Eagle's minimum essential medium (MEM) which contains essential amino acids, vitamins and salts, to McCoy's medium, which contains a larger number of different amino acids, vitamins, minerals and other extra metabolites.

(b) Bicarbonate/CO_2 buffering system

A number of supplements to the basic media are necessary to enable them to be used for culturing cells. Cell cultures have an optimum pH for growth, generally between pH 7.4–7.7. The type of buffering that is used for the media depends on the growth conditions. When cells are incubated in a CO_2 atmosphere an equilibrium is maintained between the medium and the gas phase. A bicarbonate CO_2 buffering system is most often used due to its low toxicity towards the cells. Each type of media has a recommended bicarbonate concentration and CO_2 tension to achieve the correct pH and osmolarity. Nevertheless, this may vary slightly among laboratories, therefore, a sample of media should be left under the normal incubating conditions and monitored, buffering can then be adjusted accordingly.

HEPES, another organic buffer compatible with cells, provides additional buffering capacity in the cultures and stabilizes the pH during the time that the cultures are out of the incubator and at normal atmosphere for observation and manipulation. Serum itself has considerable buffering capacity. The HEPES concentration can be reduced to 10 mM or eliminated if serum (5–15%) is to be used as a supplement.

HEPES buffer should normally be used in conjunction with bicarbonate for which a relationship between the HEPES and bicarbonate exists for differing CO_2 levels, although HEPES alone can maintain pH in the absence of exogenous CO_2. The addition of 1–5 mM pyruvate to the medium increases the endogenous cellular production of CO_2 and limits the need for a CO_2 atmosphere. Some defined media have been devised for this purpose, *e.g.* Leibovitz L-15 medium. Cells which produce large amounts of endogenous CO_2 under certain incubation conditions may require HEPES to buffer this CO_2 product. In general phenol red in the medium will indicate the state of the pH at any given time.

(c) Glutamine and amino acids

L-glutamine is an essential AA and required by virtually all mammalian cells in culture. Unlike most other amino acids, it is labile in solution and is most stable in medium stored at 4ºC. Little or no breakdown occurs until the end of the three week refrigeration. Glutamine is most stable at neutral pH (7.3) and deteriorated at acidic (3.0) and basic pH (9.0) at similar rates.

The concentration of amino acids usually limits the maximum cell concentration attainable, and the balance may influence cell survival and growth rate. Glutamine is required by most cells, although some cell lines will utilize glutamate, but evidence suggests that glutamine is also used by cultured cells as an energy and carbon source (Reitzer *et al*., 1979) in preference to glucose, although glucose is present in most defined media. In addition to buffering the medium, there are other growth requirements including amino acids, the requirement for which may vary with cell culture type. Commonly the necessary amino acids include cysteine and tyrosine, but some non-essential amino acids may be needed.

(d) Antibiotics and antimycotics

Unless good sterile conditions can be maintained (*e.g.* using biosafety cabinet, laminar flow hoods) it is necessary to incorporate antibiotics and antimycotics into the media. A wide range of suitable preparations are available from relatively specific antibiotics, *e.g.* penicillin/streptomycin, to broad spectrum antibacterial/antimycotic agents such as kanamycin or amphotericin B. The antibiotics chosen should not be toxic to the cells in culture and may depend on the type of contamination experienced in the individual laboratory. Antibiotic agents may have some toxicity and should be tested at several concentrations on the cells of interest. When preparing primary culture, an antibiotic may be added to the medium during the initial stage of tissue handling, this may be done by preparing a 1000 fold concentrated stock solution of an antibiotic such as gentamycin and adding it directly to the medium.

(e) Salts

The salts are mainly those of Na^+, K+ Mg^{2+}, Ca^{2+} Cl^-, SO_4^{2-}, PO_4^{3-}, and HCO_3^- and are the major components contributing to the osmolality of the medium. Na^+, K^+, and Cl^-, regulate membrane potential, while SO_4^{2-}, PO_4^{3-}, and HCO_3^- have roles as matrix requirements and nutritional precursors for macromolecules, as well as regulators of intracellular charge.

(f) Serum

Most cell lines require serum for adequate growth. The level of serum used depends on the particular cell line and should be determined empirically. Batches of serum may vary considerably in their ability to support cellular growth. It is therefore important to test batches of serum and have sufficient quantities of a batch that shows suitable growth supporting characteristics.

The sera used most in tissue culture are calf, fetal bovine, horse, and human. Calf serum is the most widely used, fetal bovine is second widely used, usually for more demanding cell lines and for cloning. Human serum is sometimes used in conjunction with some human cell lines. Although most cell lines still require the supplementation of the medium with serum, there are now many instances where cultures can be maintained and may proliferate on serum-free medium. Continuous cell strains such as the L929 and HeLa were among the first to be grown on serum-free media.

(g) Polypeptides

Natural clot serum stimulates cell proliferation more than serum from which the cells have been removed physically by centrifugation. This appears to be due to the release of a polypeptide from the platelets during clotting. This polypeptide is known as platelet-derived growth factor (PDGF) is one of a family of polypeptides with mitogenic activity and is probably the major growth factor in serum. PDGF stimulates growth in fibro-blasts and glia cells, but other platelet-derived factors, such as TGF may be inhibitory to growth. Heat inactivation removes complement from the serum and reduces the cytotoxic action of immunoglobulins without damaging polypeptide growth factors, but it may remove some more labile constituents and is not always as satisfactory as untreated serum.

(h) Hormones

Hormones may exhibit a variety of different effects on cells, and it is often difficult to recognize the key pathway. Insulin promotes the uptake of glucose and amino acids and may owe its mitogenic effect to this property or to activity via the IGF-1 receptor. Some growth factors (IGF-1, IGF-2) bind to the insulin receptor but also have their own specific receptors, to which insulin may bind with lower affinity. IGF-1 also stimulates glucose uptake. Growth hormone may be present in serum, particularly fetal serum, and, in conjunction with the somatomedins (IGFs), may have a mitogenic effect. Hydrocortisone is also present in serum, particularly fetal bovine serum, in varying amounts. It can promote cell attachment and cell proliferation but under certain conditions (high cell density) may be cytostatic and can induce cell differentiation.

Serum replacement experiments suggest that other hormones required for culture may be present in serum, making it necessary at present to retain serum for culture, but underlining the desirability for its replacement.

(i) Nutrients and metabolites

Serum also contains amino acids, glucose, ketoacids, lipids, such as oleic acid, ethanolamine, and phosphoethanolamine, and a number of other nutrients and intermediary metabolites. These may be important in simple media but less so in complex media, particularly those with higher amino acid concentrations and other defined supplements.

(j) Biological buffers for cell culture medium

A buffer solution resists change in hydrogen ion concentration on the addition of acid or alkali. This resistance is called the buffer action. The magnitude of the buffer action is called the buffer capacity. The buffer capacity of a particular acid and its conjugate base will be a maximum when their concentrations are equal, *i.e.* when pH = pKa of the acid, buffer capacity also depends upon the total concentrations of acid and salt as well on their relative proportion, the greater the total concentration, the greater the buffer capacity. The usual concentration of acid and salt in buffer solutions is of the order of 0.05–0.20 M and generally the mixtures have acceptable buffer capacity within the range pH ± 1. Cellular processes are sensitive to pH changes and take place in a medium, the pH of which is carefully regulated. The majority of intracellular metabolic process occurs at a pH maintained near neutrality. The sensitivity of biological processes to pH may be because it is catalyzed by hydrogen ions or it may involve a hydrogen ion as a reactant or product. Alternatively, a pH change may alter the distribution of a compound or ion across a membrane, possibly by altering the permeability of the membrane.

The criteria for buffers suitable for use in biological research should (i) posses adequate buffer capacity in the required pH range; (ii) be available in a high degree of purity; (iii) be water soluble and impermeable to biological membranes; (iv) be enzymetically and hydrolytically stable; (v) posses a pH that is minimally influenced by concentration, temperature, ionic composition or salt effect of the medium; (vi) nontoxic to the cells; (vii) only form complexes with cautions that are soluble, and (viii) does not absorb light in the visible or ultraviolet regions.

4.4 Preparation of Media

4.4.1 Balanced salt solutions

The concept of the balanced salt solution was first developed by Ringer (1886) when he discovered empirically that an excised frog's heart can function normally for long period if perfused with a solution containing NaCl, KC1, and $CaCl_2$. The importance of buffering was also recognized, and the blood buffer, $NaHCO_3$, was included for this purpose. Locke (1901) modified Ringer's salt solution to make it correspond with the higher osmotic pressure of the blood of mammals. He increased the concentration of NaCl, KC1, and $CaCl_2$, and added the organic compound glucose to serve as an energy source. Tyrode (1910) developed a Ringer-Locke's solution that contained $MgCl_2$ and lower concentrations of NaCl, KCl, and $CaCl_2$. The buffering capacity was improved by raising the $NaHCO_3$ level and adding NaH_2PO_4. Balanced salt solutions for use in tissue and cell culture were formulated over the years by various researchers Earle (1943), Hanks (1949), Dulbecco (1954), Puck (1955), and Eagle (1959). These solutions are used for a variety of purposes viz. washing cells and tissue fragments, suspending enzymes to free cells from tissues and to dislodge them from glass and plastic surfaces, and also as a basic unit in formulating cell culture media. All these salt solutions are modifications of Tyrode's, and each was formulated for a different purpose. Earle (1943) increased the concentration of bicarbonate in the salt solution to provide a high buffering capacity to neutralize the large amounts of lactic acid produced by cells growing *in vitro*. Dulbecco and Vogt (1954) used phosphate to buffer salt solution at pH 7.6. This salt solution was developed to prepare trypsin solutions for disaggregating tissues to suspensions of cells. The higher pH was used to achieve maximum trypsin activity within the physiological pH range. Eagle (1959) developed a spinner salt solution in which all the calcium was removed to minimize the clumping of cells. The phosphate concentration was increased tenfold to improve pH control, so that frequent media changes would not be needed.

A balanced salt solution (BSS) is composed of inorganic salts, usually including sodium bicarbonate. BSS is used as a diluent for more complete media, as a washing or dissection medium, for short incubations up to about 4 hr, and for a variety of other purposes that require an isotonic solution that is not necessarily nutritionally complete. The choice of BSS is dependent on the CO_2 tension. The bicarbonate concentration must be such that equilibrium is reached at pH 7.4 and 37 °C. Earle's BSS (EBSS) is commonly used as a diluent for Eagle's media MEM(S), based on Eagle's Spinner salt solution, is a variant of Eagle's (1959) minimum essential medium, deficient in Ca^{2+} to reduce cell aggregation and attachment.

Balanced salt solutions are filter-sterilized. Sterile liquid concentrates (10 ×) can also be obtained that are diluted with nine parts of sterile, distilled water before use.

4.4.2 Synthetic or defined media

Defined media vary in complexity from Eagle's MEM, which contains essential amino acids, vitamins, and salts, to complex media such as RPMI 1640, F12 and a wide range of serum-free formulations. The complex media contains a larger number of different amino acids, including nonessential amino acids and vitamins, and are often supplemented with metabolites and minerals.

4.4.3 Preparation of media for sterilization

Powdered media or salt mixtures are extremely hygroscopic and should be protected from atmospheric moisture. The entire contents of each package should be used immediately after opening. Supplements can be added prior to filtration or introduced aseptically to sterile media. The nature of the supplement will affect the storage conditions and shelf life of the medium.

Procedure:

- Take about 90% of the final volume of tissue grade water and while gently stirring add the powdered medium or salt mixture to it. Stir until completely dissolved.
- To the solution, add the required amount of sodium bicarbonate.
- While stirring, adjust the pH 0.1–0.3 units below the desired pH using 1N HCl or 1 N NaOH, which may rise during vacuum filtration.
- Add water to make up the final volume of medium.
- Sterilize immediately by filtration using a membrane with a porosity of 0.22 microns or less.
- Aseptically dispense into sterile containers. Store in dark at 2–6°C.

Preparation of autoclavable medium

Some powdered media are specially formulated to withstand the temperatures and conditions required for autoclaving.

Procedure:

- Dissolve the required amount of medium in 90% of triple distilled water.
- Add final volume of tissue culture grade water to make up the final volume, subtracting the volume of sodium bicarbonate and L-glutamine as they are added after autoclaving.
- Autoclave medium at 121°C for 15 minutes.
- Medium should be promptly removed from the autoclave to avoid extended heating and evaporation.
- Adjust the pH using sterile 1 N NaOH or 1N HCl.
- Store liquid medium refrigerated at 2–6°C and in dark.

CHAPTER 5

PREPARATION OF SERA

5.1 Introduction

Culturing of mammalian cells is usually done in a typical synthetic basal medium to meet the environmental and nutritional requirements of a given cell line. Components, such as amino acids, vitamins, inorganic salts, and a carbon source such as glucose are commonly found in the classical basal media formulation. These basal formulations normally will not support cell growth alone, but must be further supplemented with animal serum, usually Fetal Bovine Serum (FBS) at a concentration of 5–20%.

5.1.1 What is serum?

Whole blood is collected, refrigerated, and allowed to slowly clot. After centrifugation and removal of the clot, serum is collected. Serum is the blood plasma minus fibrinogens. Serum is vitally important to the growth of cells cultured in a classical basal medium. There are over 1000 different components found in serum, including proteins, electrolytes, lipids, carbohydrates, hormones, enzymes, and other miscellaneous undefined constituents. Serum also supplies growth factors, attachment factors, nutrients for proliferation and differentiation factors and inactivating toxic compounds such as proteases, free radicals etc. In addition, serum also regulates cell membrane permeability and serves as a carrier for lipids, enzymes, micronutrients and trace elements into the cell. Further albumin, and other bulk proteins found in serum minimize non-specific absorption to culture vessels and bioreactor surfaces, also influence the physical properties of the culture system, *viz*. pH, shear stress, viscosity, osmolality and gas delivery rates.

5.2 Role of Serum in Cell Culture

Fetal bovine serum, new born calf serum and serum from other adult animals are used routinely as supplements for many cell culture systems. There are several functions of serum in supporting the growth and the proliferation of different varieties of cells, in culture. Serum contains several proteins and other substances whose multifunctional components in the serum are listed below:

- Provides protease inhibitors which inactivate trypsin
- Assists in the attachment of cells to substrate
- Provides essential low molecular weight nutrients
- Provides hormones and peptide growth factors
- Membrane protectant
- Cryoprotectant

5.2.1 Serum components

Serum is an extremely complex mixture of many small and large biomolecule which provides: 1) Hormonal factors stimulating cell growth, 2) Attachment and spreading factors (biomatrix), and 3) Transport proteins carrying hormones, minerals, lipids etc. The different serum components are listed below:

5.2.1.1 *Growth factors*

Most growth factors are present in the nano and subnanogram (per ml) range in serum. Some of them are specific for cells at differentiation stages, like the haematopoietic growth factors (colony stimulating factors), some act synergistically, like epidermal growth factor (EGF) which promotes the proliferation of fibroblasts, epidermal and glial cells. Moreover, different cell types are stimulated by different growth factors *viz.* fibroblasts respond to fibroblast growth factor (FGF), EGF, platelet derived growth factor (PDGF) and somatomedins.

5.2.1.2 *Hormones*

Among the hormones, insulin is essential for the growth of nearly all cells in culture. Due to its short half life and sensitivity to inactivation by cysteine, it is usually added at a relatively high concentration. Glucocorticoids may stimulate or inhibit cells in culture depending on the cell type and cell density. They probably modulate cell proliferation by altering responsiveness to growth factors. Some cell lines require other steroid hormones *e.g.*, oestradiol, testosterone, thyroid hormones (mainly T_3).

5.2.1.3 *Attachment and spreading factors*

A major role of serum is to provide attachment and spreading factors. Mammalian cells essentially first adhere to an adequate substrate, then spread out before they start proliferating and forming a cell monolayer.

5.2.1.4 *Binding proteins*

Serum also provides proteins which carry essential low molecular mass factors. Albumin carries vitamins, lipids (fatty acids, cholesterol) and hormones among others.

5.2.1.5 *Lipids*

Serum is a rich source of various lipids required by cultured cells for survival and growth. Cell lines differ in their requirements for essential fatty acids, phospholipids, lecithin and cholesterol. For example optimum fatty acid/cholesterol/albumin balance varies radically for different haematopoietic lineages and for different maturation stages within a given lineage.

Major Serum Components Essential for Cell Survival and Growth *in vitro*

Name	*Function*
Albumin	Osmotic pressure (plasma); binding of several classes of molecules, including bile acids, fatty acids, hormones, drugs, vitamins, ions; buffering capacity
Immunoglobulin	Immune reactions; antibodies; complement activation; lymphocyte stimulation
Transferrins	Binding/transport of Fe^3
α_2-macroglobulin	Inhibits trypsin binding protein for proteases and growth factors
Haptoglobin	Binding free hemoglobin
Acid glycoprotein	Binding/transport metal ions
Fibronectin	Clot stabilization, promotes cell attachment
β_2-glycoproteins (I-II)	Complement activation
α_1-antichymotrypsin	Protease inhibitor
Growth factors like EGF, PDGF, FGF	Mitogen
Insulin	Promotes uptake of glucose and amino acids by cells
Fetuin	Enhances cell attachment
Thyroid hormones (T_3, T_4)	Oxygen consumption, energy metabolism promotes growth and differentiation of various cells
Ceruloplasmin	Copper binding
Prealbumin	Binding/transport T_3, T_4; binding retinol binding protein
Transcalciferin	Binding vitamin D and metabolites
Cl-esterase inhibitor	Complement inhibition
Cholesterol	Membrane biosynthesis
Hydrocortisone	Promotes cell attachment, induces cell differentiation
Prostaglandins	Membrane biosynthesis
Fe^{2+}, Zn^{2+}, Cu^{2+}, Mn^{+2}	Enzyme activation and others

5.2.1.6 *Minerals*

The role of various inorganic tracer elements (Cu, Zn, Co, Mn, Mo, Va, Se) present in serum has been only partially elucidated but many may act as enzyme co-factors.

5.2.2 Advantages of serum in tissue culture

Serum serves the following major functions:

1. It provides the basic nutrients for cells; the nutrients are present both in the solution as well as are bound to the proteins.
2. Serum provides several hormones, *e.g.* insulin, which is essential for growth of nearly all cells in culture, cortisone, testosterone, prostaglandin etc.
3. Several growth factors, *e.g.* platelet derived growth factor (PDGF), transforming growth factor P(TGF-p), epidermal growth factor (EGF), fibroblast growth factor (FGF), endothelial growth factor etc. present in serum are involved in growth promotion and specialized cell function. A given hormone or growth factor may stimulate growth of one cell type, while it may have no effect on another and may even be inhibitory to some other cell type.
4. A major role of serum is to supply proteins, *e.g.* fibrobnectin, which promote attachment of cells to the substrate. It also provides spreading factors that help the cells to spread out before they can begin to divide. Although cells do produce these factors, but trypsinized cells are usually unable to attach to the substrate.
5. Serum provides several binding proteins, *e.g.* albumin, transferrin, which carries other molecules into the cell. For example, albumin carries into cells lipids, vitamins, hormones etc. Transferrin usually carries iron in a non-basic form, but binding of transferrin to its receptor in cell membrane is believed to be mitogenic.
6. Serum increases the viscosity of medium and thereby, protects cells from mechanical damages, *e.g.* shear forces during agitation of suspension cultures.
7. Protease inhibitors present in the serum protect cells, especially trypsinised cells, inhibits from proteolysis.
8. The serum provides several minerals and also acts as a buffer.

5.2.3 Disadvantages of using serum in tissue culture

There are several disadvantages of using serum in the culture medium and these are:

1. Serum may inhibit growth of some cell types, *e.g.* epidermal keratinocytes.
2. Serum may contain some cytotoxic or potentially cytotoxic constituents, *e.g.* foetal calf serum contains the enzyme polyamine oxidase, which

converts polyamines like spermidine and spermine (secreted by fast growing cells) into cytotoxic polyaminoaldehydes.

3. There is a large variation in serum quality from one batch to another; which requires costly and time consuming testing for every new batch before use.
4. Some growth factors may be inadequate in serum for specific cell types and may therefore need supplementation.
5. Serum may interferes with downstream processing when cell cultures are used for production of biochemicals/biomolecules.
6. Serum protein interferers in bioassays.
7. Risk of contamination of cell cultures due to adventitious organisms (bacteria, virus, mycoplasma etc.) which might be present in the serum.

5.3 Requirement of Serum free Media

5.3.1 Serum-free cell culture

It is usually difficult to culture cells without serum. However, an increasing number of commercial companies now supply serum-free media in which a wide range of different cell types can be cultured successfully without loss of viability. Focus on Alternatives (FoA), a group of British organizations promoting alternatives to animal experiments, have been encouraging use of serum free media for cell culture. Media that are entirely free of animal ingredients are highlighted, and an index enables researchers to easily locate the most suitable products for their particular cell types (www.focusonalternatives.org.uk). The range of serum-free media available and the different cell types supported is extensive. A number of serum-free media have been designed and optimized to support specific cell types such as the commercially important cell lines CHO, Vero, MDCK and monoclonal antibody hybridomas.

5.3.2 Content of serum-free media

Serum-free media by definition lacks whole serum as an ingredient, but it may not be completely free of serum-derived products. Some commercial companies now offer a serum 'replacement' that can be added to a culture medium in place of serum. Most consist of a basal medium to which various compounds (amino acids, attachment factors, vitamins, growth factors etc.) that perform known functions are added, and these may originate from animal, human, recombinant or plant sources. For example, the protein content of some serum-free media is bovine serum albumin.

Most of these commercial serum-free media are designed primarily for hybridoma culture, but others are applicable to other cell types.

5.3.3 Potential advantages of using low serum and serum-free media

The use of low serum or serum-free media have several benefits:

- It avoid variations of serum since various batches may have variable composition and growth promoting ability
- Better reproducibility between successive cultures
- Serum cytotoxicity could be avoided
- Prevention of fibroblast overgrowth in primary culture
- Selective culture of differentiated and functional cell types from heterogeneous populations of primary cultures without preferential stimulation of variant sub populations
- Easier downstream processing and purification of culture derived products
- Prevent cruelty against animals by avoiding use serum *viz*. fetal calf serum or new born calf serum
- Less protein interference in bioassays, and reduced risk of adventitious contamination of the cell lines

5.4 General Procedure for Preparation of Sera

Serum has one of the most difficult preparation procedures in tissue culture for variations in the quality and consistency of the raw material and also because of the difficulties encountered in sterile filtration.

5.4.1 Preparation of mammalian serum

The mammalian blood is left at room temperature for an hour. The clot is removed by a glass rod and then centrifuged for 30 minutes at 3000 rpm and the serum is separated. Serum is blood plasma without any cells, platelets or clotting factors and fetal calf serum especially, is considered to be a rich source of nutrients.

5.4.2 Sterilization

Serum is usually sterilized by filtration through a 0.22-μm porosity sterilizing filter, but because of its viscosity and high particulate content, it should be passed through a graded series of cartridge type glass-fiber prefilters before passing through the final sterilizing filter. The prefilters assemblies may be stainless steel with replaceable cartridges or may be a single bonded unit.

5.4.3 Storage

Serum should be stored frozen at –20ºC. Bottle the serum in sizes that will be used up within 2–3 weeks after thawing. Freeze the serum as rapidly as possible and if thawed, do not refreeze unless further prolonged storage is required. Multiple freeze thaws should be avoided because it may cause denaturation of proteins. Serum is best used within 6–12 months of preparation if stored at –20ºC, but more prolonged storage may be possible at –70ºC. Do not store serum in a frost free freezer, since temperature cycling may cause bottles to crack and may contribute to deterioration of product. Polycarbonate or high density polypropylene bottles will eliminate the risk of breakage if stored at –70ºC.

5.4.4 Serum thawing

Animal sera are the most important source of nutrients for cultured cells. They contain lipids, salts, vitamins, minerals, amino acids and other necessary components for cell growth. In order to preserve these properties for long periods, careful handling is crucial. The serum bottle is removed from the freezer and left to stand at room temperature for 10 minutes. The bottle is then placed in a 37ºC water bath, with the bottle cap clear of the water (not submerged). The bottle is shaken occasionally until the serum is completely thawed.

5.4.5 Quality control

Before going for a new serum lot, a sample of the new lot should be tested for quality control assays *viz.* Cell plating efficiency assay, Cloning efficiency, Growth curve and Batch reservation.

5.5 Contamination of Serum

Serum is a major constituent of cell culture media which is always derived from some animal source *e.g.* fetus of newly born animals or adult animals. One disadvantage of using animal sourced biological fluids, such as fetal bovine serum (FBS) is the possible introduction of adventitious microbial contamination. Sometimes, these contamination levels are very low. A variety of methods have been applied to reduce or eliminate contaminating viruses in animal derived materials including FBS. The methods include heat treatment, pH, ultracentrifugation, chemical inactivation and irradiation. Gamma irradiation has been found to be the most effective inactivation process which is not detrimental to the performance of the treated product.

Some infectious agents found in bovine serum are: Bovine adenoviruses Type I, V, Bovine herpes virus-l (BHV-l), Infectious bovine rhinotracheitis (IBR) virus, Bovinc parvo virus, Bovine entero viruses, Bovine viral diarrhoea (BVD) virus, Para influenza type-3 (PI-3) virus, Rabies, Reovirus, Bacteriaphage spp, Mycoplasma spp., Enterobacteria spp. etc. Similarly reported infectious agents in human serum are: Hepatitis-B virus and infectious Cytomegalo virus (CMV), Hepatitis-C, D and E viruses, Epstein barr virus-2 (EBV-2), B-Lymphotropic herpes virus (HBLV) and Human T-Iymphotropic virus (HTLVs).

CHAPTER 6

PRIMARY CULTURES

6.1 Introduction

Primary culture is defined as isolation of the cells from a tissue, but before the first subculture. Three stages are involved in the process (i) isolation of the tissue, (ii) dissection and release of cells, and (iii) culture of cells by seeding into the culture vessel. Following isolation, a primary cell culture may be obtained either by allowing cells to migrate out from tissue, adhering to a suitable substrate by disaggregting mechanically or enzymatically. For most normal untransformed cells it appears to be essential to attach to a flat surface in order to survive and proliferate. Transformed cells on the other hand, particularly cells from transplantable animal tumors, often grow in suspension. The enzymes used most frequently in these cultures are crude preparations of trypsin, elastase, collagenase, DNase, hyaluronidase either used alone or in combinations. The purified enzymes are generally less toxic and more specific, example trypsin and pronase. Hyaluronidase can be used in conjunction with collagenase to digest the intracellular matrix, and DNase is employed to disperse DNA released from lysed cells as it tends to impair proteolysis and promote re-aggregation. Although each tissue may require a different set of conditions, certain requirements are shared by most primary cultures:

- Fat and necrotic tissue should be removed by dissection.
- The tissue should be chopped finely with sharp blade.
- Enzymes used for disaggregation of cells, should be removed.
- The concentration of cells in the primary culture is generally kept much higher.

- A rich medium is preferable to a simple medium.
- Specific cell types require selected media.
- Embryonic tissue yield better results than adult tissue.
- Fetal bovine serum may be used to give better survival of cells.

6.2 Source of Material

Primary cell cultures can be obtained from any source of tissue. The condition of the cells and their behaviour in culture will always be affected by the material chosen for generation of culture. Before choosing the source it is important to decide, (i) whether to use normal or tumour derived tissue, (ii) adult or embryonic tissue, and (iii) species of origin.

6.2.1 Adult or embryonic tissue

Cultures can be derived from adult tissue or from embryonic tissue. Cultures derived from embryonic tissue generally survive and grow better than those taken from adult tissue. Tissues from almost all parts of the embryo are easy to culture, whereas tissues from adult are often difficult to culture. This presumably reflects the lower level of specialization and presence of replicating precursor or stem cells in the embryo. Adult tissues will usually have a lower growth fraction and a high proportion of non-replicating specialized cells, often within a more structured, and less readily disaggregated, extra cellular matrix. Initiation and propagation are more difficult, and the life span of the culture is often shorter. Embryonic or fetal tissue has many practical advantages, but it must always be remembered that in some instances the cells will be different from adult cells and it cannot be assumed that they will mature into adult-type cells unless this can be confirmed by appropriate characterization. Examples of widely used embryonic cell lines are various mouse embryo fibroblasts and MRC-5 and other human fetal lung fibroblasts.

Embryonic stem cells

A more recent development has been the removal of embryonic stem cells (ES-cells) from the embryo during the blastocyst stage of development. These cells can be grown in culture for many generations in their undifferentiated state and are of particular interest because they can be manipulated in culture and then re-introduced into embryos.

Normal or neo-plastic tissue

Normal tissue usually gives rise to cultures with a finite lifespan. These cells will generally grow as an undifferentiated stem cells or precursor cells and the onset of differentiation is accompanied by a cessation in cell proliferation which may be permanent. Cultures from tumors can give continuous cell lines, although there are several examples of continuous cell lines (3T3 fibroblasts) which are non-tumorigenic. Tumor tissue can often be passaged in the syngeneic host, providing a cheap and simple method of producing large numbers of cells.

6.2.2 Adherent or suspension culture

Cells may grow as an adherent monolayer or in suspension. Adherent cells are said to be anchorage-dependent and attachment to a substratum is a prerequisite for proliferation. They are generally subjected to contact inhibition, which means they grow as an adherent monolayer and stop dividing when they reach such a density that they touch each other. Most cells, with the exception of mature hemopoietic cells and transformed cells, grow in this way. In contrast to anchorage dependent cells, cells cultured from blood, spleen or bone marrow adhere poorly to the culture dish. In the body, these cells are held in suspension or are only loosely adherent. The methods used to propagate these cells are very different to those for adherent cells. Suspension cultures are easier to propagate, since subculture only requires dilution with medium. Cultures in which cells grow attached to each other or to a substratum have to be treated by a proteolytic enzyme to break the bond between cells and substratum. The most commonly used enzyme is trypsin while other enzymes, *viz.* collagenase, papain, dispase, and pronase, are also used.

6.2.3 Primary cultures/continuous cell lines

If you remove tissue from an embryo, dissociate it into a suspension of single cells, and plate them out onto a culture dish, a series of characteristic events occurs. Firstly, cells are in a lag phase, usually no more than 1–2 days in length, during which there is little or no increase in cell number. During this time, cells are "conditioning" the medium, undergoing internal cytoskeletal and enzyme changes and adjusting to the new medium. Secondly, the cells undergo a period of rapid division, called log phase of growth. Then, as they approach confluency and form contacts with one another, their rate of division slows and they begin to express a program of differentiation characteristic of their tissue of origin. Muscle cells fuse and acquire cross-striation, while heart cells begin to beat spontaneously.

Primary cultures described as above, are prepared from cells taken directly from the animal. The cells divide, acquire differentiated characteristics, and ultimately die. For the next batch of primary culture, it is back to the animal again to obtain new tissue and prepare new culture. Alternatively, in the case of cells that divide in culture, it is possible to 'passage' or 'subculture' them by inducing them to detach from the substrate, 'splitting' them and allowing them to re-enter log phase growth.

However, the properties of the cultured cells often change gradually with passaging, as more rapidly dividing cell populations come to predominate and more 'differentiated' cells, those divide more slowly, are lost. When cells are repeatedly sub-cultured, most cease division after a finite number of generations, typically between 20 and 80. This is thought to reflect the same process of senescence that occurs in cells *in situ.*

It is possible however, to develop populations of cells that can be passaged indefinitely and that express a reasonably stable phenotype. These are referred to as established or continuous cell lines. Some cell lines have arisen spontaneously in normal cells being passaged in culture, but the majority has been obtained by culturing tumor cells. In addition to their infinite life span or 'immortality', such cell lines frequently share several additional properties that distinguish them from 'normal' cells in culture. They divide more rapidly, do not require attachment to the substratum for growth, and when reintroduced into animals, they form tumors. Cell lines with these properties are also referred to as transformed cell lines. Once a continuous cell line has been established, it is customary to clone the cells in order to obtain a genetically homogeneous population. Many separate clonal cell lines, each exhibiting a unique set of properties, can be obtained from a single tumor, or tumor cell line.

6.2.4 Primary cultures: Advantages and limitations

As previously described, a primary culture is defined as one started from cells, tissues or organs taken directly from organisms. The major advantages of primary cultures are the retention of (a) The capacity for biotransformation; and (b). Tissue specific functions. In many cases, the metabolic profile generated by a primary cell culture such as hepatocytes has greater similarity to *in vivo* than does the pattern seen with subcellular fractions used as an exogenous source for biotransformation. The second advantage of primary cultures is the retention of tissue specific functions. One limitation of primary cultures is the necessity to isolate cells for each experiment. Procedures to isolate cells require the disruption of the tissue, often with proteolytic enzymes. This may result in the loss or damage of specific membrane receptors, damage to the integrity of the membrane, and loss of cellular products. Primary cultures have a limited life span and changes in metabolism and tissue specific functions will occur with time in culture. In summary, primary cultures offer a cell system which can often provide a better approximation of the intact tissue than cell lines.

6.2.5 Continuous cell lines: Advantages and limitations

The following important considerations should be taken into account before undertaking any experimental study with a cell line:

1. Before use, cultures should be carefully examined for homogeneity of cell type, and if necessary, must be re-cloned.
2. Population doubling time may also affect the number of cells seeded in order to obtain a required density.
3. Type of cell and culture conditions used (*i.e.*, suspension versus adherent cultures; sparsely seeded versus dense cultures).
4. Information concerning normal or tumorigenic characteristics should be noted. This may also be reflected in chromosome number and ability to replicate in medium with low serum concentration.
5. Some cell lines maintain or secrete tissue-specific proteins and enzymes which can be used for examination as biological markers. Specific biomarkers should be noted, such as special receptor sites or responses to specific growth factors.
6. Temperature effects may yield important information which can be obtained from cultures adapted to lower temperature or cell lines.
7. Cell lines must be clean of contaminating mycoplasma, bacteria or viruses.
8. Frequent monitoring of stock cultures is mandatory to detect changes in growth patterns and morphology.
9. Experiments may be repeated several times, using the same environmental conditions to determine variability.

6.3 Processing of Tissue for Primary Culture

On the basis of the use and function of particular cell type, selection of cells is to be done example normal cell or neoplastic cells, anchorage dependent cell or suspension cultures, embryonic stem cells or adult stem cells. Cells are isolated from appropriate tissue origin depending on its properties and use.

Blood cells which are present in the circulation can easily be obtained. The problem is of separating the cell type of interest if the cells of interest are organized into solid tissue, *e.g.* kidney or skin, then it is required to disaggregate the cells and purify the cell type of interest. The choice of method for disaggregating the cells depends both on the nature of the tissue and the amount of material available. Best results are obtained with fresh tissue. The tissue should be washed free of blood and the fat and damaged tissue removed prior to isolation of the cells.

6.3.1 Isolation of chicken embryo fibroblasts (CEF)

Fertile 10-days incubated eggs are cleaned with 70% (v/v) ethanol; the broad end of the egg is opened with the help of forceps and the embryo is poured into a sterile petri-dish. Using a pair of sterile forceps, the head is chopped off and the trunk transferred to a fresh petri-dish containing 5 ml of sterile calcium and magnesium free Hank's balanced salt solution (CMFH). Wash the embryos twice in sterile CMFH at room temperature (25°C) to remove the erythrocytes.

Lay the embryo on its back, use sterile curved forceps to pinch out the internal organs and breast tissue. Discard the internal organs and the embryo is transfered to a dry sterile culture dish. The embryo is chopped into small pieces with sterile scissors. Resultant tissues should be suspended in CMFH and kept on ice. The pieces can now be transferred to a sterile conical glass flask containing glass beads. 25 ml of cold 0.2% (v/v) trypsin (1:250) should be added and the flask swirled at room temperature for 30 sec. After allowing the larger fragments to sediment, discard the supernatant once. Further 15–20 ml of trypsin should be added, swirled at room temperature for 30 sec and the liquid phase transfered to a sterile container. This should be repeated four to five times, centrifuged for 5 min at 4°C and 300 g, the supernatant discarded and the cells re-suspended in 10 ml CMFH containing 5% (v/v) fetal calf serum (FCS) The cell suspension should now be poured through a cell strainer of 220 μm mesh The cells should be pelleted as before and re-suspended in 20–30 ml of DMEM containing 10% (v/v) FCS. The cells can now be counted and plated in cell culture flasks. Change the medium to remove any dead cells, and if necessary sub-cultured by trypsinization. These cells may grow for up to 20 passages; however, some specialized strains will grow for only a week before becoming senescent.

6.3.2 Separation of mononuclear cells from peripheral blood

Collect blood in a heparinized syringe or vacutainer. Place 10ml blood onto 10ml of (pre-warmed) lymphocyte separation medium in a 50 ml centrifuge tubes. Centrifuge at 400 g for 30 min at room temperature. Use a Pasteur pipette to collect the opaque interface containing the mononuclear cells. Transfer the opaque interface to a clean (sterile) 15 ml centrifuge tube. Wash the cells 3–4 times with 30 ml phosphate buffered salien (PBS) or medium without serum and centrifuge at 250–300 g for 5–10 min. Wash the cells one more time, then re-suspend in an appropriate culture medium, *e.g.* RPMI 1640 with 10% FCS.

6.3.3 Preparation of murine spleen cells

Using sterile scissors and forceps cut open the abdomen and remove spleen carefully and transfer to sterile PBS. The spleen is pocked at multiple sites with needle and then PBS is flushed in the spleen with force many times until a fine cell suspension is obtained. Transfer spleen cell suspension to 15 ml centrifuge tubes and centrifuge at 100 g for 30 sec. Collect the supernatant to a fresh 15 ml tube and centrifuge at 300 g for 5 min. Re-suspend the cell pellet in RPMI 1640 with10% FCS and plate the cells.

6.3.4 Mouse embryo cell culture

A convenient source of cells for undifferentiated fibroblastic cultures is from mouse embryos. They are often used as feeder layers. Remove uterus aseptically from 13 days pregnant mouse when the embryo is relatively large having large proportion of undifferentiated mesenchyme. Dissect out the embryos by incising the uterus and transfer the embryos to a fresh dish and wash with sterile DBSS. The embryos are chopped and treated with trypsin for few hours, dissociated cells are collected every half an hour on ice. Check and count the viable cells with haemocytometer. Dilute the cells to 1×10^6 in growth medium and seed the flasks with 2×10^5 cells/cm^2.

6.3.5 Primary culture of lamb testis (LT)

Healthy, prepubertal (45–60 days old), male lambs, apparently free from any infectious disease, is a good source of testicles for primary LT cell culture. Bilateral orchiectomy (open method) under local anesthesia may be carried out for the collection of testicles. The testicles were collected in chilled PBS under sterile conditions. All cell culture work is performed in clean air work station. The testicles are thoroughly washed in chilled PBS supplemented with antibiotics. Tunica vaginalis, albugenia and eqididymis were removed and the resultant testicular tissue is washed, minced thoroughly and finally washed with chilled PBS. The supernatant is carefully decanted off.

Minced testicular tissue is transferred into a trypsinization flask containing 50–60 ml of 0.25% trypsin, along with a sterile teflon coated magnetic bar. Trypsinization is carried out at 37ºC for 10–15 min by constant stirring and the supernatant is carefully decanted into a sterile flask containing 20–25 ml GMEM supplemented with 10% NBCS. The resultant suspension is stored at 4ºC. Trypsinization process is repeated 4–5 times so as to accomplish maximum yield of cells.

The pooled cell suspension is filtered through a sterile double folded muslin cloth and the filtrate is centrifuged at 1000–1200 rpm for 10 min, the supernatant is discarded and the cell pellet is re-suspended in 10ml medium, pooled and recentrifuged. The final pellet is homogeneously mixed with growth medium by gentle pipetting. Cell density is adjusted to 0.5–10^6 cells/ml in culture flasks and incubated at 37ºC. Cells are examined daily to assess the growth pattern and monolayer formation.

CHAPTER 7

CELL LINES

7.1 Introduction

A wide range of cultured animal cell lines has been described in the literature and are available from tissue culture collections such as the European Collection of Cell Cultures (ECACC) in the UK, the American Type Culture Collection (ATCC) in the USA and Indian Type Culture at Pune. These cell lines are from different animal species, including humans. Such lines, while providing a valuable resource and being appropriate for many uses, are not suitable for all purposes. In this situation the scientist may consider establishing his or her own primary culture and, if necessary, try to isolate a new cell line. This chapter describes the approaches which are available and to indicate, wherever possible, how to go about deciding which method to use.

7.2 Cell Line

A cell line is continuous or finite depending on whether it has limited culture life span or it is immortal. On the basis of the life span of culture, the cell lines are categorized into two types:

(a) ***Finite cell lines*****:** The cell lines which have a limited life span and go through a limited number of cell generations (usually 20–80 population doublings) are known as Finite cell lines. These cell lines exhibit the property of contact inhibition, density limitation and anchorage dependence. The growth rate is slow and doubling time is around 24–96 hr.

(b) ***Continuous cell lines*****:** Cell lines transformed under laboratory conditions or in vitro culture conditions give rise to continuous cell lines. The cell lines show the property of ploidy (aneupliody or heteroploidy), absence of contact inhibition and anchorage dependence. They grow in monolayer or suspension form. The growth rate is rapid and doubling time is 12–24 hr.

Majority of continuous cell lines grow as monolayers. Some of the cells which are non-adhesive *e.g.* cells of leukemia or certain cells which can be mechanically kept in suspension, can be propagated in suspension. There are certain advantages in propagation of cells by suspension culture method.

These advantages are:

(a) The process of propagation is much faster.,
(b) The frequent replacement of the medium is not required.,
(c) Suspension cultures have a short lag period,
(d) Treatment with trypsin is not required,
(e) A homogenous suspension of cells is obtained,
(f) The maintenance of suspension cultures is easy and bulk production of the cells is easily achieved.,
(g) Scale-up is also very convenient.

The cell lines are known by: (a) A code *e.g.* NHB for Normal Human Brain. (b) A cell line number- when several cell lines are derived from the same cell culture source, *viz*: NHB1, NHB2. (c) Number of population doublings, the cell line has already undergone *e.g.* NHB2/2 means two doublings.

Continuous cell lines

Some of the properties exhibited by continuous cell lines are:

1. ***Tumourogenecity*:** Continuous cell lines are tumorogenic *in vivo* and by testing it can be observed that whether the cell line is capable to form a tumor or not.
2. ***Differentiated status*:** Most cell lines stop expressing tissue specific genes but retain expression of house keeping genes which are required for continued growth in culture as a result of which most of the cell lines are phenotypically more alike compared to the original tissue from which the cell line was derived.
3. ***Heterogeneity and instability*:** Continuous cell lines show phenotypic heterogeneity and genotypic variation.
4. ***Neuploidy*:** There is a considerable variation in both number of chromosomes and type of specific chromosomes.

Infinite cell lines exhibit some features such as:

1. Can be cultured continuously with infinite life span.
2. Reduced requirement for serum or growth factors.
3. Shorter population doubling time.
4. Reduced substrate adhesion.
5. Genetic instability.
6. Altered growth control.
7. Loss of contact inhibition.
8. Loss of anchorage dependence.
9. Loss of density dependent inhibition.

7.3 Culturing Continuous Cell Lines

Conditioned medium

Conditioned medium is a medium which has already been used to support the growth of cells. Although the original cells will have depleted it of components such as glucose and glutamine, it will still contain growth factors and hormones secreted by these cells. It can, therefore, be used to stimulate the growth of some cells. Conditioned medium is generally prepared by removing medium which has been in contact with cells in logarithmic growth for about three days and centrifuging it to remove cells, followed by filtration through a 0.22 µm filter to remove micro organisms and cells. The medium is stored at –20°C and used at around 10–25% of the new medium.

Feeder cells

Feeder cells are often used to condition the culture medium and stimulate cell growth, in particular at low density. Cells, such as irradiated or mitomycin C-treated fibroblasts, peripheral blood lymphocytes, or peritoneal exudate cells, provide an undefined mixture of substances which promote cell growth, and may be the only way of obtaining growth at low cell density.

Normal 'versus' tumour

Many of the existing cell lines have been derived from tumour tissue, and this is particularly true of human cell lines where tumour tissue is often easier to obtain and easier to culture than normal material. The choice of tumour tissue may give important benefits if the cell type of interest is rare in the normal tissue but present in large quantities in an individual with a cancer of that cell type, *e.g.* a particular stage of lymphocyte differentiation. The advantage of normal tissue is, of course, that there is less concern about pathological changes present when the culture is established.

Adult 'versus' embryo

It is generally accepted that, in the case of tissue which is capable of cell division in culture, the potential number of cell doublings *in vitro* correlates well with the age of the tissue at isolation. Thus one can expect the maximum number of cell divisions to be obtained from embryonic tissue. The disadvantage of embryonic material is that in some tissues, *e.g.* liver, there are substantial phenotypic differences between fetal and adult material. Also, in the case of human tissue, there are ethical issues associated with use.

Anchorage dependent cells

Following culture sometimes cultured cells exhibit a character like anchorage dependence. The cells require attachment to a surface in order to survive, to proliferate or express their differentiation function. However, non-anchorage dependent cell require no solid substrate and they have a capacity to proliferate in a fluid suspension or within a semisolid medium.

Culture of adherent cells

Culture of attached cells involves the growth of cell on a rigid substrate. Cells are seeded in a liquid medium into a culture flask or dish. Cell attaches and proliferates and can be sub cultured. Routine culture substrate for an anchorage dependent cell includes glass, plastic or various types of porous filter membrane. The most common is glass culture vessels, *viz.* petri-dishes, culture flask, and roux flask. Glass is rarely used for mass culturing, while glass cover slips and slides give the best image by light microscopy and phase constraint microscopy. Polystyrene is the most commonly used plastic because of excellent clarity. Many cell types will not adhere with the unmodified plastic.Therefore pre treatment making the surface more hydrophilic is required.

Defects of cell attachment and growth

Improper media volume viz. too low or high volume may affect cell survival. Also insufficient mixing can influence the pattern of cell attachment. It will set the cell unevenly more concentration towards the center and less concentration towards the periphery. Different cells bind with different intensity and they also differ in how tightly they adhere to the substrate and therefore peeling off more tightly bounded cells results in their destruction. Vibration can also affect cell attachment, excessive vibrations can delay attachment and can aggregate cells to a specific area. Temperature gradient can also affect cell growth. Attachment of source cells may be sensitive to temperature and when the temperature within incubator is not uniform it results in either a heavy growth or poor proliferation.

Enhancement in anchorage of cells by use of attachement factors and biosubstrate take place in the sysetm as attachment factors are agents that promote cell adhesion to the conventional culture substrate. They can increase cell spreading and cell proliferation. Attachment factors include:

- Basic polymers like polylysines which increase the surface charge.
- Purified cell components like fibronectin, collagens and laminin.

Adherent cells can be cultured on 3D matrices including collagen sponge, plasma clot, fibric clot, medium hybrated collagen gels. Collagen gels are versatile and can be used in the study of cell-cell, cell-matrix interaction, cell mobility and for the growth of tumour cells.

Adherence cells can also be cultured on the micro carries with a surface suitable for cell attachment which can increase surface area: volume ratio. They are available in various sizes and are made up of dextric, collagen, plastic, gelatin, polystyrene, cellulose, plastic coated glass or glass coated plastic.

CHAPTER 8

CELL COUNTING AND STAINING

8.1 Introduction

Cell viability in culture is measured by various techniques. Most of these are based on loss of membrane integrity which is generally detected by the uptake of dyes. Cells die via two distinct ways, apoptosis and necrosis. Different techniques are used for staning apoptotic cells, as the cell membrane does not disintegrate.

Changes in culture conditions, such as accumulation of metabolic wastes, serum depletion, changes in the pH or temperature leads to cell death. Cells undergoing apoptosis usually shrink followed by nuclear condensation, nuclear fragmentation, and degradation of the DNA into nucleosomal fragments. Cell death can be induced by several agents/factors. It has been demonstrated that a wide variety of stimuli such as removal of growth factors from the culture, heat shock, radiation, various genotoxic drugs and cytotoxic agents all induce cell death. All these agents kills cells via apoptosis, are capable of inducing both apoptosis and necrosis in a dose-and time-dependent fashion.

Nature of the assays

(a) ***Staining*****:** Fix cells in a 100% methanol and immerse the slide in 70% methanol solution, once the slide is fixed, stain the nucleus. This is done by dipping the slide in a solution of acid dye (0.1% w/v Eosin Y, 0.1% w/v formaldehyde, 0.4% w/v sodium phosphate dibasic). Then stain the slide for cytoplasm by dipping the slide in solution of basic dye (0.4% w/v methylene blue-polychrome, 0.4% w/v azure, 0.4% w/v sodium phosphate dibasic, 0.5% w/v potassium phosphate monobasic). Wash the slides with water, air-dry, mount and view under microscope.

(b) *Viability*: Viability assays can be used to measure the proportion of viable cells following primary disaggregation, cell separation, or freezing and thawing. Most viability tests rely on a breakdown in membrane integrity that is determined by the uptake of a dye to which the cell is normally impermeable (*e.g.* trypan blue, erythrosin or naphthalene black).

(c) *Irritancy*: A response analogous to inflammation allergy, or irritation *in vivo*, may be possible to assay by monitoring cytokine release in organotypic cultures.

(d) *Transformation*: Survival in an altered state *e.g.* state expressing genetic mutation or malignancy or transformation.

(e) *Metabolic*: Microtitration based assays are followed to measure metabolic response *e.g.* dehydrogenase activity, DNA, RNA, or protein synthesis

(f) *Survival*: The long-term retention of self-renewal capacity which may be 5–10 generations or more.

8.2 Packed Cell Volume

The packed cell volume (PCV) is the proportion of red blood cells in blood volume. The packed cell volume (PCV) can be determined by centrifuging heparinized blood in a capillary tube (also known as a microhematocrit tube) at 10000 rpm for 5 minutes. This separates the blood into layers. The volume of packed red blood cells, divided by the total volume of the blood sample gives the PCV. As a capitulary tube is used for this calculation can be done by measuring the length of the cell layers.

8.3 Haemocytometer Counting

The easiest, direct and simplest method of counting cells in suspension is haemocytometer counting. By this method percentage of viable cells can be determined using the dye exclusion method. The haemocytometer is a modified microscope slide which displays a precisely ruled, subdivided grid. The grid consists of nine primary squares, each measuring 1 mm on a side and limited by three closely spaced lines (2.5 µm apart). These triple lines determine if cells lie within or outside the grid. Each of the primary squares is further divided to make cell counting easier. The plane of the grid rests 0.1 mm below two ridges that support a sturdy cover slip. There is depression at the edge of polished surface. The cell suspension is added at the edge of side to be drawn across the grid by capillary action.

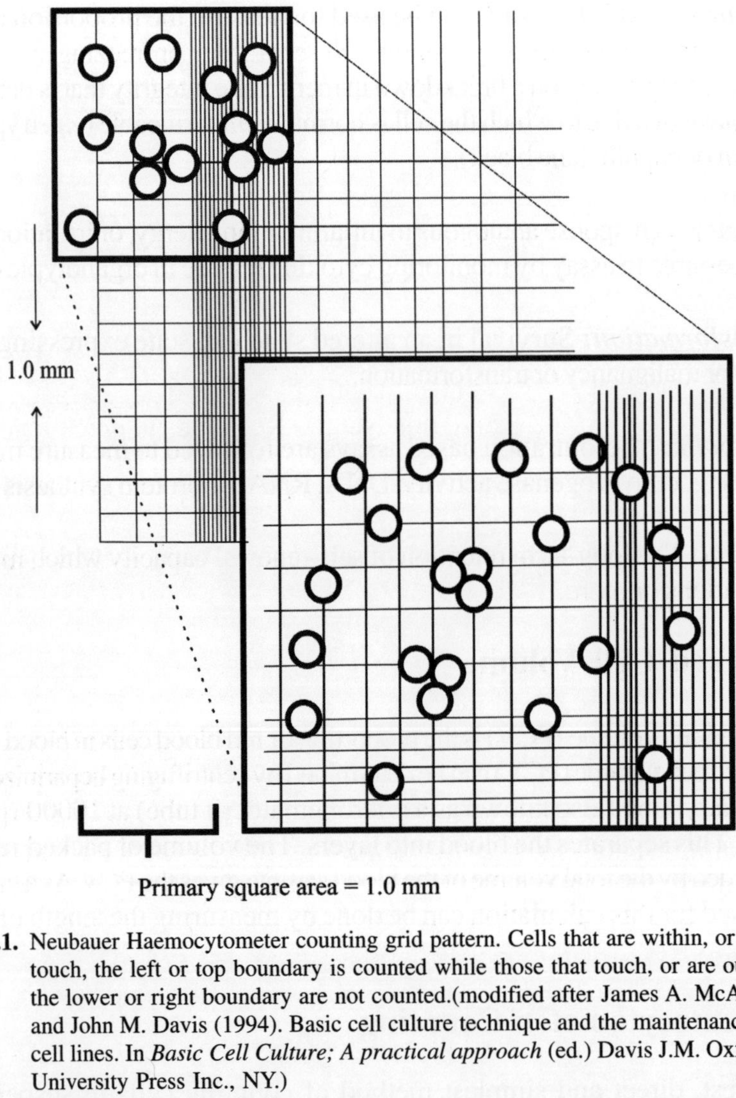

Fig. 8.1. Neubauer Haemocytometer counting grid pattern. Cells that are within, or that touch, the left or top boundary is counted while those that touch, or are outside, the lower or right boundary are not counted.(modified after James A. McAteer and John M. Davis (1994). Basic cell culture technique and the maintenance of cell lines. In *Basic Cell Culture; A practical approach* (ed.) Davis J.M. Oxford University Press Inc., NY.)

When performing cell counts with a hemocytometer for accuracy and reproducibility, counts must be performed in the same way each time.

Cell count = average cell count of five primary square $\times 10^{-4} \times$ dilution factor

This gives total cell concentration in the original suspension in cells/ml. Care should be taken that the cell suspension is uniform and ideally without cell clumps. Clumping can be minimized by diluting the sample. For viable cell counts, allow a set time of 4–5 min for dye to diffuse into the cells, as too rapid a count may underestimate the number of damaged cells.

8.4 Electronic Particle Counting

Cells in suspension can be counted accuratcly and rapidly using a particle Coulter counter (Beckman Coulter). Many instruments are available that allow cell sizing and the determination of size distribution within a cell suspension. Electronic particle counters operate on the principle that when cells are suspended in an electrolyte and a metered volume of this suspension is passed through a narrow (20–200 μ.m) aperture that carries a nominal current, as a cell crosses the orifice it produces an increase in impedance resulting in a voltage pulse. The pulse is amplified, and the number of pulses within the set threshold is counted and displayed on an oscilloscope. Cell sizing can also be performed, as pulse amplitude is directly proportional to cell volume. Cell clusters or debris may also register a pulse, but these extraneous counts can be eliminated by adjusting the threshold of the pulse amplitude.

8.5 Estimation of Viability

Trypan blue exclusion assay

The live cells do not take up trypan blue, this fact is exploited for cell viability assay. This assay allows distinguishing between cells with intact and disrupted membrane. Using a hemocytometer, mix an equal volume of cells with an equal volume of a 0.2% solution of trypan blue prepared in 2 × PBS and observe cells under microscope for ascertaining live and dead cell count. Calculate the number of cells/ml, in culture medium by multiplying the average number of cells per field by the dilution factor of 2 and hemocytometer index (10^4).

The percent non-viability is determined as:

% Dead cell = number of stained cells/total number of cells × 100

And percentage viability is determined as:

% Viable cell = number of unstained cells/total number of cells × 100

Dye uptake assay

Fluorescein diacetate is the dye which is taken by viable cell and hydrolysed to fluorescein which in turn is impermeable to cell membrane. Fluorescein is green in colour hence live cells appear fluorescent green. Non viable cells may be stained with propidium iodide and subsequently floresce red. This method may be applied to fluorescene microscopy or flow cytometry by adding the dye mixture at a proportion of 1:10 to give a final concentration of 1μg/ml of diacetyl fluorescein and 50 μg/ml propidium iodide. Incubate for 10 min at 37°C. Finally place a drop of stained cells on a slide, and examine under fluorescent microscope. Cells that fluoresce green are viable, while those that fluoresce red are non viable. Percentage of viable and nonviable cells can be calculated as given above.

8.6 Neutral Red Staining

Neutral Red is a eurhodin dye which stains lysosomes in red. It can be used as a vital stain, to stain living cells. Neutral red is not retained by dead cells. Uptake of neutral red is quantified by fixing the cells in formaldehyde and solubilising the stain in acetic ethanol, allowing the plate to be read on an ELISA plate reader at 570 nm. Neutral Red tends to precipitate, so the medium with stain is usually incubated overnight and centrifuged before use. This assay does not measure the total number of cells, but it does show a reduction in the absorbance related to loss of viable cells. It is used to stain cell cultures for plate titration of viruses.

CHAPTER 9

CHROMOSOME KARYOTYPING

9.1 Introduction

The number and appearance of complete set of chromosomes in the nucleus of a eukaryotic cell in a species is termed as Karyotype. The length, the position of the centromeres, or any other difference between the chromosomes is characteristic of the species and is an extremely valuable research tool for the one who works with cell cultures. There can be karyotypic variations from line to line in established cell lines derived from the same species, or from cell to cell within a line. This genetic heterogeneity should be properly known in order to properly interpret and evaluate karyotype information from cell culture data. When tissues are explanted and cultured, the cells that grow generally display the karyotype of the individual. However, when cell cultures become established the rigid control over the karyotype is generally lost, and chromosomal variation becomes common. As there is variation in chromosome number, an established cell line is usually characterized by its modal number rather than its absolute number of chromosomes. In general, when the modal number is close to the normal diploid number, it is often termed as pseudo diploid or quasi diploid cell line. When the modal number is higher it is a heteroploid cell line. Other than numerical variation, many cell lines undergo extensive chromosomal rearrangement like deletion, translocation, inversion, etc. resulting in new "marker" chromosomes. Six different characteristics of karyotypes are usually observed and compared.

1. Differences in absolute sizes of chromosomes. Chromosomes can vary in absolute size by many folds between genera of the same family. This feature probably reflects different amounts of DNA duplication.
2. Differences in the position of centromeres.

3. Differences in relative size of chromosomes can only be caused by segmental interchange of unequal lengths.
4. Differences in basic number of chromosomes may occur due to successive unequal translocations.
5. Differences in degree and distribution of heterochromatic regions. Heterochromatin stains darker than euchromatin, indicating tighter packing, which mainly consists of genetically inactive repetitive DNA sequences.
6. Differences in number and position of satellites, which are small bodies attached to a chromosome by a thin thread.

A full account of a karyotype may therefore include the number, type, shape and banding of the chromosomes, as well as other cytogenetic information.

9.2 Preparation of Karyotype

Cultured lymphocytes, long-term lymphoblast cultures or any cell line that grows in suspension can be a sample for karyotyping. Any monolayer culture also can be trypsinized and then handled as a suspension culture. Other technique is an *in situ* method whereby monolayer cultures are processed right on their growth surface in specially designed culture chambers. The cells are taken, cultured and stained. Metaphase spreads are selected, and the chromosomes are arranged in descending order by size and compared with a standard. Both size and staining pattern are included in the analysis. There are basically five steps in the preparation of slides for Karyotyping.

(a) Growth optimization

For cells in culture, a high mitotic index is usually easy to achieve. Growth of the cells is optimized, for some cell lines, changing the medium before making chromosome preparations may be helpful. Rapidly growing lines are usually sub cultured 2 days before and slower growing lines 3–4 days before "harvesting" the cultures for chromosome studies.

(b) Metaphase arrest

Colchicine or colcemid are two commonly used chemicals for blocking cells in metaphase. Exposure time of the cells to these chemicals depends on the length of the cell cycle. Faster growing cells are exposed for shorter time where as longer treatments are required for slower growing cells. Higher concentrations can cause excessive contraction of the chromosomes loosing the clarity of banding patterns.

(c) Hypotonic swelling

Swelling of the cells is a critical step and is essential to achieve well-spread chromosomes. A variety of hypotonic solutions are in use, but 0.075 M potassium chloride is most popular and is a good choice. Treatment time varies (15–45 min) for different cell lines.

(d) Fixation

Swelling always follows fixation of cells. The universal fixative is mixing methanol and glacial acetic acid in 3:1 ratio. It is usually made fresh before use and kept cold until use. It is always added slowly to prevent cells from clumping. To ensure complete fixation the fixative is changed 3–4 times before drying the cells on a slide.

(e) Air drying

Optimum spreading of the cells depends on drying of the fixative which in turn depends on the temperature and relative humidity in the laboratory. Drying can be achieved by blowing, fanning, placing the slide on a warm surface or hot plate.

9.3 Staining

Staining of chromosomes is done for exploiting their morphological differentiation. Earlier, chromosomes could be stained uniformly over their length. Since then, several procedures have been developed for differentially staining certain regions of the chromosomes to give them a banded appearance. Like C-banding, R-banding, Q-banding or quinacrine banding, in which chromosomes are stained with quinacrine resulting in patterns of alternating bright and dull fluorescence along the chromosome; G-banding or Giemsa banding, a technique in which chromosomes are treated with trypsin followed by staining with Giemsa producing a series of dark staining.

Staining techniques

(a) Solid staining

Giemsa stain is the most commonly used stain for uniform staining of the chromosomes. For solid staining, 3 coplin jars or 50 ml beakers are used. In first beaker 50 ml of Gurr's buffer (pH 6.8) or 10 mM phosphate at pH 6.8 containing 1.5 ml of concentrated Giemsa, in the second and third beaker 50ml of water in each is placed. Place slides into the stain for 3–4 min and rinse by dipping 8–10 times in each beaker of water. Shake off excess water and allow to dry on a paper sheet to drain off and dry.

(b) Giemsa (G-) banding

Giemsa can be used far solid staining as well as in other banding procedures. In the original G-banding technique the critical step was incubation of the slide at high temperature. This step has been replaced by incubation of the slide in a crude trypsin solution. The Giemsa stain is the same as that used for solid staining. As for solid staining, the Gurr's buffer in which the stain is diluted can be re-placed with 10 mM phosphate at pH 6.8.

A prerequisite for good G-banding is a high-quality spread on slide. Chromosomes should be straight and slender with sister chromatids lying close together. Cells should be well spread and thus many mitotic cells would be included on a slide. These properties can be checked by phase-contrast microscopy before proceeding with banding. Following either solid staining or G-banding, slides can be studied microscopically.

Types of banding

G-banding is obtained with Giemsa stain following digestion of chromosomes with trypsin. It yields a series of lightly and darkly stained bands- the dark regions tend to be heterochromatic, late-replicating and AT rich, while light regions tend to be euchromatic, and GC rich.

R-banding is the reverse of G-banding (R stands for "reverse"). The dark regions are euchromatic (G-C rich regions) and the bright regions are heterochromatic (A-T rich regions).

C-banding Giemsa binds to constitutive heterochromatin, so it stains centromeres.

Q-banding is a fluorescent pattern obtained using quinacrine for staining. The pattern of bands is very similar to that seen in G-banding.

T-banding is to visualize telomeres.

Spectral karyotype (SKY technique)

Spectral karyotyping is a molecular cytogenetic technique used to simultaneously visualize all the pairs of chromosomes in an organism in different colors. Fluorescently-labeled probes for each chromosome are made by labeling chromosome-specific DNA with different fluorophores. Because there are a limited number of spectrally-distinct fluorophores, a combinatorial labeling method is used to gencratc many diffcrcnt colors. Imagc proccssing softwarc assigns a pscudo color to each spectrally different combination, allowing the visualization of the individually colored chromosomes. This technique is used to identify structural chromosome aberrations when Giemsa banding or other techniques are not accurate enough to analyse.

Chromosome abnormalities

Chromosome abnormalities can be numerical, as with the presence of extra or missing chromosomes or structural as in translocations, inversions, large-scale deletions or duplications. Numerical abnormalities are also known as aneuploidy often occur as a result of non-disjunction during meiosis in the formation of a gamete; trisomies, in which three copies of a chromosome are present instead of the usual two, are common numerical abnormalities. Structural abnormalities often arise from errors in homologous recombination. Both types of abnormalities can occur in gametes and therefore will be present in all cells of an affected person's body or they can occur during mitosis and give rise to a genetic mosaic individual who has some normal and some abnormal cells.

Uses of karyotyping

(a) Gene dosage

In the studies on regulation of enzyme synthesis, it is desirable to know whether the cell line is diploid, pseudo diploid or heteroploid.

(b) Identification of cell lines

Karyotype of the cell line are usually characteristic of the line, and three or four such unique markers can provide an unequivocal identification test for a cell line.

(c) Monitoring of cell lines

Experiments involving contact with cell lines and subcellular preparations such as chromosome transfer experiments carry the danger of cross contamination by live cells from the donor line. To avoid cross contamination of cell lines, karyotype analysis is used for monitoring the purity of cell lines.

(d) Verification of hybrid formation

In interspecies hybrids the chromosomes provide a definitive set of markers to confirm that the putative hybrid is in fact a hybrid of the desired type. In intraspecies hybrids, the use of parental cell lines with different characteristic marker chromosomes can often be used for verifying that the hybrid is of the desired one.

In the work on gene mapping, karyotyping plays a major role. Several techniques have been developed for the same. Some of these include:

1. C-banding, a procedure for specifically staining the centromere regions of chromosomes as well as regions of constitutive heterochromatin.
2. A technique employing incorporation of bromodeoxyuridine (BUdR) into DNA (in place of thymidine) allowing one to differentially stain and therefore to distinguish chromatin with one BUdR substituted strand from that with 0 or 2 substituted strands; the method has been used to study replication kinetics of the individual chromosome bands and to differentially stain sister chromatids of the same chromosome allowing quantitation of sister chromatid exchanges.
3. Nucleolus organizing region (NOR) staining is used to specifically stain the nucleolus organizing regions of chromosomes containing the sites of the 18 S and 28 S ribosomal RNA genes.
4. Cytological hybridization, a procedure in which radioactively labeled RNA or DNA is hybridized to denatured chromosomes followed by autoradiography to identify the sites of complementary chromosomal DNA.

CHAPTER 10

GROWTH KINETICS

10.1 Introduction

Growth is an irreversible increase in size, mass or number. In case of cells it is increase in number and size with time. There are various phenomena related to growth. Many growth phenomena in nature exhibit a logarithmic or exponential increase. The size, mass or number increases by a constant, similar to simple compound interest. Thus growth occurs at a compounded rate or called as exponential growth.

10.2 Absolute Growth Rate

When growth is plotted (size, mass or number) versus time (Fig. 10.1), a constantly increasing growth curve is obtained. The slope between any two time points, gives the absolute growth rate(AGR), which is the change in actual growth over time.

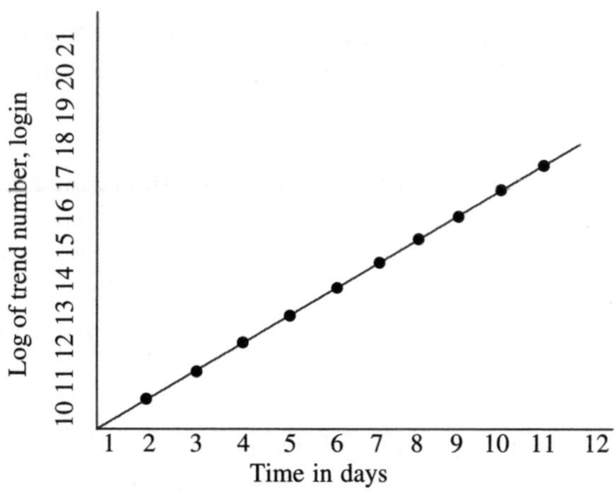

Fig. 10.1. Absolute growth rate (AGR)

10.3 Relative Growth Rate

When logarithm of growth (size, mass or number) versus time (Fig. 10.2), is plotted a linear line is obtained. The slope of the line, gives the relative growth rate, which is the change in relative growth over time. Since the line is linear, relative growth rate is same regardless of any time interval is chosen to calculate the slope.

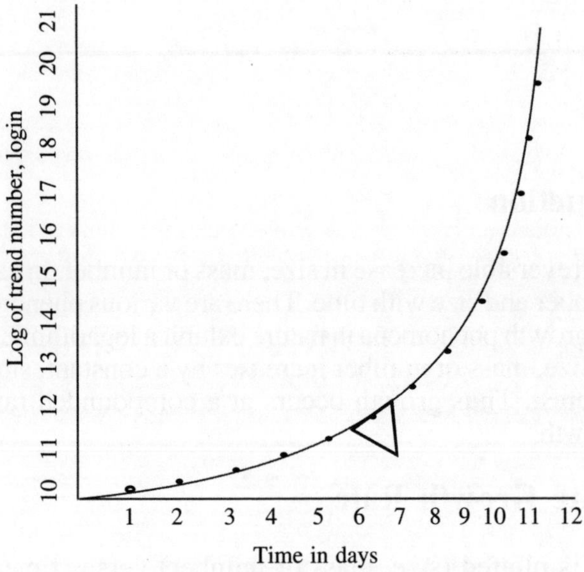

Fig. 10.2. Relative growth rate (RGR)

Cells growth progresses through a characteristic growth pattern which consists of lag phase, log phase (or exponential phase) and the stationary (or plateau) phase. The log and plateau phases are important and provide vital information about the cell line.

***The Lag Phase*:** This is the time immediately following subculture and during this phase there is no increase in the cell number. During this period the cell replaces the cell surface elements and extra cellular matrix lost during trypsinization. Cells adhere to the substrate and start spreading out. During this phase the synthesis of new DNA and structural proteins takes place and cytoskeleton reappears, which is an integral part of the spreading process.

***The Log Phase*:** This is the phase which follow the lag period. Here exponential increase in the cell number follow doubling the population and terminate after confluence is reached. The length of the log phase depends on the seeding density and the growth rate of the type of cells. In the log phase the population is at its most uniform state and the viability is also high.

***The Plateau Phase*:** When the culture becomes confluent, and the cells are in contact with surrounding cells, this marks the end of log phase. The growth rate of the culture is now reduced, and in some cases cell proliferation ceases almost completely. At this stage, the culture enters the plateau, or stationary phase. Some fibroblasts become oriented with respect to one another, forming a typical array of cells.

Contact inhibition phenomena was originally described by Abercrombie and Heaysman (1953) which describe the cessation of motility, membrane ruffling, and slowing growth following contact of cells at confluence. The reduction in the growth of normal cells after confluence is due to contact inhibition and depletion of nutrition and growth factors in the medium. Cultures of normal simple epithelial and endothelial cells stop growing after reaching confluence and remain as a monolayer. Cultures that have transformed spontaneously or have been transformed by virus or chemical carcinogens will usually reach a higher cell density in the plateau phase than their normal counterparts. This higher cell density is accompanied by a higher growth fraction and the loss of density limitation of cell proliferation. The plateau phase for these cultures is equilibrium between cell proliferation and cell loss. These cultures are often anchorage independent and do not require any substrate to grow.

Direct and indirect methods to measure cell mass

Methods for measurement of the cell mass involve both direct and indirect techniques.

1. Physical measurement of dry weight, wet weight or volume of cells after centrifugation. This is the direct method.
2. Chemical measurement of some chemical component of the cells such as total nitrogen, total protein, or total DNA content, as direct method.
3. Chemical activity such as measurement of production or consumption of gases like CO_2, or O_2 gives an indirect assessment of cell mass.
4. Turbidity measurements employ amount of light scattered by a suspension of cells, particulate objects such as bacterial cells scatter light which is proportional to the number of cells present. The turbidity or optical density of a cell suspension is directly related to cell mass.

10.4 Determination of Doubling Time

The number of generations that occur per unit time is multiplication rate (r) and is usually expressed as population doublings per 24 hr.

Multiplication rate = Number of generations / unit time (day)

Population doubling time (PDT) is the time, taken for cell number to double, it is expressed in hr, and is the reciprocal of the multiplication rate (*i.e.* 1/r).

PDT = total time elapsed/number of generations

The growth characteristics of the cell population as a whole are described by population doubling level, multiplication rate and population doubling time. These however, do not characterize the division cycle of individual cells. By definition PDT (commonly 15–25 hr) is not the same as cell generation time (cell cycle time), which is the interval between successive divisions for an individual cell. In practice, however PDT is used as an estimate of cell cycle time and to determine the length of the phases of the cell cycle.

Before calculation of PDT, cells must be sub-cultured at regular and fixed intervals and when the culture is in the log phase of growth. Each subculture should be at the same seeding density and culture conditions must be consistent, including the same type of substrate, size of flask, volume of medium, and formulation of medium. The method used to perform cell counts must be consistent and accurate.

Calculation of the growth rate constant

During the exponential (or logarithmic) growth phase, a culture mimics a first order chemical reaction, *i.e.* the rate of increase of cells is proportional to the number of cells present at that time. The constant of proportionality μ is an index of the growth rate and is called the growth rate constant:

Rate of increase of cells = μ × number of cells

The value of μ can be determined from the following equation:

$$\ln N_t - \ln N_0 = \mu (t - t_0)$$

N_t is natural log of the number of cells at time **t**

N_0 is the number of generation at zero time

The mean generation time or "doubling time" (g) is the average time required for all the components of the culture to double. This is calculated from the equation:
$\log_{10} N_t = \log_{10} N_0 + g \log_{10}{}^2$

10.5 Cell Differentiation and Cell Cycle

10.5.1 Cell differentiation

Rate of cell proliferation are usually measured by the response of cells to a particular stimulus of culture conditions. Testing medium, serum, new culture flasks, all require quantitative assessment of culture growth. Estimation of culture growth is done in routine maintenance for knowing the best time to subculture, the optimum dilution, and the estimated plating efficiency at different cell density. Details of the growth kinetic parameters are critical in the design of cell culture experiments as cultures vary significantly in lag phase, log phase and plateau phase. Therefore the status of the culture, the duration of the population doubling time (PDT) and the cell cycle time should be known. Generally cell cultures are consistently uniform in the log phase and sampling at the end of the log phase gives the highest yield and greatest reproducibility. It is also important to consider the transition from one state to another. Cells that have entered the plateau phase have different morphology, they are more differentiated, and may become polarized.

10.5.2 Cell cycle

Phases of the cell cycle

There are four phases of cell cycle G_1, S, G_2, and M. The total length of the cell cycle can be determined with the length of each of the successive phases of the cell cycle (G_1, S, G_2, and M). The commonly applied methods are based on detection of DNA synthesis or on quantification of DNA content. Cells synthesize DNA only during the S phase (DNA synthesis phase) of the cell cycle, so incorporation of the radiolabelled nucleotide ^{3}H thymidine into DNA is limited to this phase only. Cells that incorporate label can be identified by autoradiography.

10.5.2.1 *Labelled mitoses method*

The length of the G_1, S, G_2, and M phases of the entire cell cycle can be obtained by determining the percentage of labelled cells in mitosis. The method includes; multiple replicate cultures are pulsed with [^{3}H] thymidine (370 Bq/ml) for 20–30 min, unincorporated [^{3}H] thymidine is washed away, and incubated. Samples of three cultures are then fixed at successive time (0, 2, 4, 30 hr) and processed for autoradiography. After (^{3}H) thymidine pulse the percentage of labeled mitoses is determined and plotted versus time. The length of the complete cell cycle is 15 hr for most cultured animal cells. Therefore, if the percentage of labelled mitoses is determined at points up to 30–35 hr the data will include some cells that have cycled through two rounds of mitosis. The plot of percent labelled mitoses versus time shows a rapid rise (first rise) to almost 100%, a fall (first fall) to a few percent, and then another rise (second rise) as some cells enter their second round of mitosis.

Total cell cycle time (S + G_2+ M + G_1)

(i) *S phase*: The period of DNA synthesis is the time interval between the 50% labelling points on the first rise and the first fall which is usually 7–8 hr. (ii) *G_2 phase (post-DNA synthesis, pre-mitosis gap phase)*: This is the time between the end of the [^{3}H] thymidine pulse and the 50% labelling point on the first rise (usually 2–4 hr). (iii) *Length of mitosis*: This is not evident from the curve, but is approximately the same for all normal cells (usually 0.5–1 hr) and (iv) *G_1 phase*: The post-mitotic gap which precedes DNA synthesis is estimated as the time between the 50% point on the first fall and the 50% point on the second rise. G_1 is the most variable phase of the cell cycle (*e.g.* it may not exist for some rapidly dividing cells), but is usually 6–15 hr.

Some workers estimate the interval between successive mitoses using the time between any two corresponding points for percent labelled mitosis on the first rise (first mitosis) and the second rise (second mitosis).

Cell cycle analysis by DNA content (Flow cytometry)

The amount of DNA within a cell changes as the cell progress through the phases of the cell cycle. DNA content is greatest during the G_2 and M phases and lowest during post-mitotic G_1 gap, and increases throughout S phase. Flow-cytometry of a dye-treated (certain dyes binds to DNA) cell suspension can be used to obtain a population distribution based on relative DNA content. The percentage of cells in each phase is directly proportional to the length of that phase.

10.6 Plating Efficiency

Plating efficiency reveals differences in the growth rate within a cell population and distinguishes between alterations in the growth rate (colony size) and cell survival (colony number). Cells grow as isolated colonies at low cell densities. Differential clonal growth rates reflect differences in the capacity for cell proliferation between lineages within a population. These differences are not however expressed in an interacting monolayer at `higher densities when cell communication is possible. When cells are plated out as a single-cell suspension at low cell densities (2–50 cells/cm^2), they grow as discrete colonies. The number of these colonies can be used to express the plating efficiency.

CHAPTER 11

CULTURE PARAMETERS

11.0 Introduction

Monitoring the life cycle of a culture effectively necessitates acquaintance with culture parameters which need to be kept in mind while maintaining healthy cultures. The various culture parameters are described below:

11.1 pH

Most cell lines grow at pH 7.4. The optimum pH for cell growth varies relatively little among different cell lines, some normal fibroblast lines perform best at pH 7.0–7.7, and transformed cells do better at pH 7.0–7.4. In general it may be advantageous to do a brief growth experiment to determine the optimum pH. Phenol red is generally used as an indicator which is red at pH 7.4 and becomes orange at pH 7.0, yellow at pH 6.5, lemon yellow below pH 6.5. In alkaline medium, it becomes pink at pH 7.6, and purple at pH 7.8. Since the assessment of color is highly subjective, it is useful to make up a set of standards using a sterile balanced salt solution (BSS) and phenol red at the correct concentration to arrive at the best results.

11.1.1 CO_2 and bicarbonate

In the medium, carbon dioxide remains in dissolved state and is in equilibrium with bicarbonate to maintain the pH. The atmospheric CO_2 tension will directly regulate the concentration of dissolved CO_2 as a function of temperature. This in turn produces H_2CO_3 which dissociates into H ion and bicarbonate ion as shown in the reaction.

$$H_2O + CO_2 \text{-} \longleftrightarrow H_2CO_3 \longleftrightarrow H^+ + HCO_3\text{-} \quad (A)$$

Since HCO_3 is unstable, it tends to re-associate, leaving the medium acidic. The net result of increase in atmospheric carbon dioxide is to lower the pH. So the effect of elevated carbon dioxide tension is neutralized by increasing the bicarbonate concentration

$$NaHCO_3 \longleftrightarrow Na- + HCO_3- \quad (B)$$

The increased HCO_3 concentration pushes equation (A) to the left until equilibrium is reached at pH 7.4. If another alkali (*e.g.* NaOH) is used instead, the net result is the same:

$$NaOH + H_2CO_3 \longleftrightarrow NaHCO_3 + H_2O$$
$$NaHCO_3 \longleftrightarrow Na+ +HCO_3 + H_2O$$

When preparing new medium for the first time, add a specified amount of bicarbonate and then sufficient 1N NaOH such that the medium equilibrates to the desired pH after incubation at 37°C overnight. Each medium has a recommended bicarbonate concentration and carbon dioxide tension for achieving the correct pH and osmolarity, but minor variations will occur in different methods of preparation. With the introduction of HEPES buffers into tissue culture, it was thought that, since CO_2 was no longer necessary to stabilize the pH, it could be omitted. This proved to be untrue, at least for a large number of cell types, particularly at low cell concentrations. Although 20 mM HEPES can control pH within the physiological range, the absence of atmospheric carbon dioxide allows equation to move to the left, eventually eliminating dissolved CO_2, and ultimately HCO_3, from the medium.

When pyruvate is added to the medium, it enables cells to increase their endogenous production of CO_2. This makes them independent of exogenous source of CO_2 as well as HCO_3. Higher concentration of sodium pyruvate, (550 mg/l), in Leibovitz L15 medium does not require CO_2 in the gas phase and also lacks $NaHCO_3$. L15 is sometimes recommended for the collection of biopsy samples as it does not require CO_2. Required buffering is provided by high amino acid concentration. Sodium-β-glycerophosphate can also be used to buffer autoclavable media lacking CO_2 and HCO_3. If the limitation of CO_2 is their then the above may be used but only after appropriate testing.

In short, cultures in open vessels need to be incubated in carbon dioxide atmosphere which is in equilibrium with the sodium bicarbonate in the medium. Cells at mildly high concentrations that is more than 10^5 cells/ml are grown in sealed flasks with no carbon dioxide added to the gas phase, provided that the bicarbonate concentration is kept low (~4 mM), particularly if the cells are high acid producers. At low cell concentrations, however (*e.g.* cloning), it is necessary to leave the cap slightly loose in a CO_2 incubator.

11.1.2 Buffering

Buffering of culture media is done for open dishes, where the evolution of CO_2 causes the pH to rise, and in case of transformed cell lines where high cell concentrations causes overproduction of CO_2 and lactic acid resulting in fall of pH. A buffer may be incorporated into the medium to stabilize the pH, but exogenous CO_2 may still be required by some cell lines, particularly at low cell concentrations, to prevent the total loss of dissolved CO_2 and bicarbonate from the medium. In spite of its poor buffering capacity at physiological pH, bicarbonate buffer is still preferred over others, because of its low toxicity, low cost, and nutritional benefit to the cultures. In the pH 7.2–7.6 range, HEPES is much stronger buffer and is used frequently at 10 or 20 mM. It has been found that, when HEPES is used with exogenous CO_2, the HEPES concentration must be more than double that of the bicarbonate for adequate buffering. A variation of Ham's F12 with 20 mM HEPES, 10 mM bicarbonate, and 2% CO_2, has been used successfully in many laboratories for the culture of a number of different cell lines. It allows handling of terasaki plates and other multi-well plates without an excessive rise in pH and minimizes the requirement for HEPES.

11.2 Oxygen

Oxygen is very important for cell culture. Hemoglobin analoges have been made to incorporate oxygen carriers, and cells rely chiefly on dissolved oxygen, which can be toxic due to the elevation in the level of free radicals. Providing the correct oxygen tension is, therefore, always a compromise between fulfilling the respiratory requirement and avoiding toxicity. Strategies involving both elevated and reduced oxygen levels have been employed, as has the incorporation of free radical scavengers, such as glutathione, 2-mercaptoethanol or dithiothreitol, into the medium.

Most dispersed cell cultures prefer lower oxygen tensions, and some systems *e.g.* human tumor cells do better in less than the normal level of atmospheric oxygen tension. The depth of the culture medium can influence the rate of oxygen diffusion to the cells, it is advisable to keep the depth of the medium within the range 2–5 mm (0.2–0.5 ml/cm^2) in static culture.

11.3 Osmolality

Most cultured cells have a fairly wide tolerance for osmotic pressure which may be different for different species *viz.* ~310 mOsm/kg for mice and ~290 mOsm/kg for human. In general practice, osmolalities around 260 mOsm/kg are quite acceptable for most cells, but, should be kept consistent at ± 10 mOsm/kg. To compensate for evaporation during incubation slightly hypotonic medium may be better for petri-dish or open plate cultures. The measurement of osmolality is a useful quality control step in making up the medium, as it helps to check errors in weighing, dilution, etc. It is particularly important to monitor osmolality if alterations are made in the constitution of the medium.

11.4 Temperature

The optimal temperature for cell culture is dependent on the body temperature of the species from which the cells were obtained. The temperature recommended for most human and warm blooded animal cell lines is 37ºC close to body temperature. Because of the higher body temperature in birds, avian cells may be maintained at 38.5ºC for maximum growth, but will grow satisfactorily at 37ºC.

Temperature set should be a little lower for safety, as overheating is a serious problem than under heating. Cultured cells will tolerate considerable drops in temperature, and can survive for several days at 4ºC, and can be preserved at –196ºC, but they cannot tolerate more than 2ºC above normal temperature for more than a few hours and will die quite rapidly at 40ºC and above.

To ensure reproducible results attention must be paid to the consistency of the temperature (within ± 0.5ºC). Doors of incubators or hot rooms must not be left open for long. The spatial distribution of temperature within the incubator or hot room must be uniform without any “cold spot” within, and air should circulate freely. Space must be allowed between flasks for air to circulate in the incubator or hot room.

Cold-blooded animals that do not regulate their blood heat within narrow limits tolerate a wide temperature range, between 15ºC and 26ºC. So for simulating *in vivo* conditions an incubator may require with cooling as well as heating aid to keep the temperature below ambient.

11.5 Viscosity

The serum content in the medium mainly influences the viscosity of a culture. Viscosity of the medium in most cases will have little effect on cell growth. However, it is important when a cell suspension is stirred or when cells are dissociated after trypsinization, any cell damage that occurs under these conditions may be reduced by increasing the viscosity of the medium with carboxymethylcellulose (CMC) or polyvinylpyrrolidone (PVP).

11.6 Surface Aeration

In a closed system of cell culture, such as a sealed flask, one of the important factor is the amount of oxygen in the flask and the availability of this oxygen to the cells growing under 3–6 mm of medium. In order to provide sufficient oxygen to the growing cells in a flask normally a ratio of 10:1 between head space and medium volume is used. Thus 1 litre flask (*e.g.* a Roux bottle) with 900 ml of air and 100 ml of medium will initially contain 0.27 g oxygen which will support cells for satisfactory growth and maintenance for quite long period. The transfer rate of oxygen from the gas phase into a liquid phase has been calculated at about 17 μg/cm^2/h. Again, this is well in excess of that required by cells in a 1 litre flask. These calculations show the importance of maintaining a large head space volume for good areation, otherwise oxygen limitation could become one of the growth-limiting factors in static closed cultures.

11.6.1 Sparging in bioreactors

Very effective way of transferring oxygen to culture is through bubbling the culture. Usually large air bubbles with very low gassing rate is used. Sparging is used in a culture vessel having great height diameter ratio to increase the efficiency of gas. The narrow vessels with more height creates higher pressure at the base of the vessel which in turn increases oxygen solubility.

11.7 Cell Density

The cell density can be quantified as grams of dry/wet weight per litre of sample, or as number of viable/dead cells per ml. The cells in a sample can be separated by centrifuging and weighed while they are wet, or the cells may be thoroughly dried before weighing. The dry weight measurement usually gives a much more consistent result than the wet weight. Alternatively, the number of cells can be counted either by successively diluting the original sample and plating on a petri-dish, with the help of a microscope and a counting chamber, or with an automated cell counter such as a Coulter counter or a flowcytometer. The plating method detects only the viable cells; whereas, the automated cell counters can detect the total number of cells. Practically the most often used method simply monitors the optical density of the sample. The absorbance of the sample measured in a spectrophotometer is correlated to either the dry weight or the number of cells per volume.

Biomass concentration is one of the most critically needed measurements in fermentation studies. It is also one of the most difficult and unreliable one. Similarly, the optical density measurement only has limited usefulness if the fermentation broth is not clear. In addition, these methods cannot distinguish the viable cells from the dead ones. On the other hand, the standard plate count can detect viable cells among other particulate matters. However, the method requires elaborate preparation and is costly. Densitometry has been used to quantify the density of cell growth within a culture vessel. The cells are fixed then stained with 0.5% crystal violet and the light transmittance (or absorbance) of the specimen is determined. The method is most useful in detecting the relative differences in culture density has been applied to growth factor response studies in primary culture and to assays of colony formation used to access nutrient requirements for optimal cell growth.

Cell density of a given sample can be measured with the following five methods: wet weight, dry weight, optical density, direct cell counting with a chamber, and successive dilutions followed by plating.

Dry/wet weight measurement

1. Pour out 100 ml of the culture into a graduated cylinder after stirring the flask to suspend the culture evenly.
2. Separate the cells from the broth by centrifugation at 10000 g for 5 minutes and scrape the cell paste from the centrifuge tube into a weighing pan. In the case of filtration, the culture is poured into the holding reservoir fitted on the filter membrane. A vacuum is applied to pull the liquid through the membrane. The wet weight of the culture is measured immediately after all the water has been pulled through.
3. Dry the cell paste in an oven set at 100ºC. Measure the weight of the pan/filter plus the cell paste periodically until there is no further decrease in the dry weight. It will take 6–24 hr to dry the sample completely. Calculate the difference in the weight, and express the dry weight in g/l.

Optical density

1. Dilute the sample to appropriate concentrations, and measure the absorbance of the sample with a spectrophotometer at 550 nm or other wavelengths. The accuracy of the method is the highest when the absorbance is between 0.1 and 0.5.
2. The optical density also depends on cell morphology, size and shape, as the amount of transmitted or scattered light depends strongly on these factors. For a given culture sample, a linear relationship between the number of cells and the absorbance may be obtained. As a thumb rule, an optical density of 1 unit corresponds to approximately 1 g/l of dry cell.

Cell counting chamber

This is a standard method of counting which is widely used in blood cell counts also commonly called hemocytometer. While loading the sample should not overflow nor it be too less to cover the area under coverslip. Count the average number of cells in each square and then multiply this number with 10^4 and dilution of the cells giving number of cells per ml. Two thin rails of a well defined height of 0.02 mm are attached to the surface of a glass slide that is marked with evenly spaced lines at 0.05 mm intervals. Thus, each square represents a volume of 5.0×10^8/ml.

Standard plate count

The culture is serially diluted with sterile water. The serially diluted culture is poured into petri-dishes along with the nutrient agar. The number of colonies is counted after incubation at 37°C.

1. Shake the flask containing the culture, and pipette out 1 ml of the culture in sterile bottle, which contains 99 ml of sterile water. Mark it as number "1". The dilution factor in bottle 1 is 1:100.
2. Using another sterile pipette, transfer 1 ml from the bottle "1" into a bottle marked "2", which also contains 99 ml of sterile water. Shake and mix. The dilution factor in this "bottle 2" is 1:10000.
3. With the sterile pipette, transfer 1 ml from the bottle 2 to a petri-dish marked 1:10000. Thus, the number of colonies from this dish is multiplied by 10^4 to give the number of cells in 1 ml of the original sample.
4. Using the same pipette transfer 0.1 ml of the sample making serial dilution up to a final cell count of 10^7 in 1 ml of the original sample.
5. Heat 50 ml of agar to boil for 10 minutes. Cool the agar to 45°C, pour 12 ml to each petri-dishes. The culture should not be either too hot (it will kill the culture), or too cold otherwise it will solidify before pouring.
6. Swirl the plates gently but thoroughly to mix the culture with the agar. Now allow the agar to solidify.
7. Incubate the petri dishes at 37°C for 48 hr.
8. Every large and small colony is counted and marked with permanent marker pen. A cell counting chamber can be used in conjunction with cell plating to distinguish the number of viable and nonviable cells.

CHAPTER 12

SCALING UP OF CULTURE SYSTEMS

Animal cell culture is scaled-up either by batch culture or continuous culture. Batch culture is used when a definite number of cells are required and cells are harvested and processed as a batch before all nutrients are used up. While continuous culture is a technique used to grow cells in a particular phase of growth. Continuous culture system can keep the cellular content homogenous for longer period of time. Batch culture is a closed culture system while continuous culture system is an open culture system.

Laboratory procedures are normally scaled up via intermediate models of increasing size and volume. The first pre-requisite for any large scale cell culture system and its scaling up is the establishment of a cell bank. Master cell banks (MCB) are first established and they are used to develop Master Working Cell Banks (MWCB). The MWCB should be sufficient to feed the production system at a particular scale for the predicted life of the product. The cell stability is an important criteria so MWCB needs to be repeatedly sub-cultured and each generation should be checked for changes. The volume of cultured cells is maintained by i) increasing the culture volume, and ii) by increasing the concentration of cells in a reactor by continuous perfusion of fresh medium, so that the cells keep on increasing in number without the dilution of the medium.

A fully automated bioreactor maintains the physicochemical and biological factors to optimum level and maintains the cells in suspension medium. The most suitable bioreactor used is a compact-loop bioreactor. The animal cells grow very slowly. The main carbon and energy sources are glucose and glutamine. Lactate and ammonia are their metabolic products that affect growth and productivity of cells. Thus, monitoring of glucose, glutamate, and ammonia is carried out by on line flow injection analysis (FIA) using gas chromatography (GC), high performance liquid chromatography (HPLC) etc.

12.1 Batch Culture

Batch culture is a large-scale closed culture system in which cells are grown in a fixed volume of culture medium under specific environmental conditions (*e.g.* nutrient type, temperature, pressure, aeration etc.) up to a certain density.

In batch cultures, mainly Roller Bottles with Micro Carrier Beads (for adherent cells) and spinner flasks (for suspension cultures) are used in Scale-up of animal cell culture process.

Different methods are used for the scale up of anchorage dependent cells. These are cell factories, roller bottles and microcarriers. When scaling up anchorage dependent cells the aim is to maximize the available surface area for growth and minimize the volume of medium and headspace

Roller bottles

The Roller bottles provide total curved surface area for cell growth. Nearly all the internal surface in a roller bottle is available for cell growth, however, only 15–20% is covered in medium at any one time. The continuous rotation of the bottles in the CO_2 incubators helps to provide medium to the entire cell monolayer in culture. By rotating the bottle (0.5–5 rpm) the cells are subjected to medium and air alternatively. Less medium is required with this technique, however, headspace volume must be left to supply adequate oxygen and pH levels. The attachments rotate the bottles along the long axis which helps to expose the entire cell monolayer to the medium during one full rotation. This system has the advantage over the static monolayer culture: (a) it provides increase in the surface area, (b) provides constant gentle agitation of the medium, (c) provides increased ratio of surface area of medium to its volume, which allows gas exchange at an increased rate through the thin film of the medium over the cells.

Micro carrier beads

Micro carrier beads are small spherical particles with diameter 90-300 micrometers, made up of dextran, glass, cellulose, gelatin or collagen. By using microcarrier technology, small spherical beads act as supports for the cells while the cells grow in a suspension with gentle agitation. Micro Carrier beads, increase the number of adherent cells per flask. These beads come in a range of densities and sizes. The cells grow at a very high density which rapidly exhausts the medium and therefore the medium has to be replaced for the optimum cell growth. This technology also improves control of culture parameters, increases production yield, reduces requirement for serum and thus lowers the risk of contamination.

Suspension cells

The use of suspension cultures is preferred method for the scale-up of mammalian cells. These cultures do not require an attachment surface to facilitate their growth, therefore they can be described as 'anchorage independent' cultures. Many cells perform best in suspension cultures. The most common methods for initiating the scale-up of suspension cultures are the use of shake flasks and spinner flasks, with varying degrees of agitation, before being introduced to bioreactors capable of culturing larger volumes (ranging from 100–10000 liters). Anchorage independent cultures are very adaptable. However, it is possible to persuade many anchorage dependent cell lines to grow in suspension. Generating such cell lines from an anchorage dependent origin can be effected through two methods viz: selection and adaptation.

Static suspension cultures

Static suspensions are cultured using the same method for the culturing of anchorage dependent cells, as discussed above. As the name suggests, no agitation or mixing is involved in growing static suspension cultures. Numerous cell lines, including lymphocytic cell lines have been developed using static suspension cultures. This system follows the underlying principle for the culturing of monolayer cells, which is disadvantageous for scaling up and therefore static suspension cultures may be deemed incompatible for large scale production.

12.2 Scale-up Technology and Considerations

On scaling up of animal cell culture processes, a wide range of parameters must be evaluated. Development and optimization of parameters are essential for successful scale-up. Optimising large scale processes are difficult because the conditions within the vessels are of a more heterogeneous nature then those experienced in the smaller vessels. Many considerations associated with scaling up of mammalian cell cultures arise from the difficulty of ensuring a homogeneous environment to be maintained within the large systems. Both the physical and chemical parameters related to scale-up pose significant problems.

The whole scale up process, from a small vial of cells right up to large scale bioreactors, the whole process from the start to finish may take approximately three months or more which is depicted below:

Animal cell culture 1 ml vial

↓

Added to 75 ml of growth media in a 500 ml roller bottle

↓

Cells grow to the appropriate cell number (1×10^6)

↓

Cells divided into 500 ml roller bottles (4 nos)

↓

All are pooled into 10 L bioreactor

↓

10 L perfusion cultures, which are harvested daily (3 nos)

↓

All pooled into 100 L Bioreactor

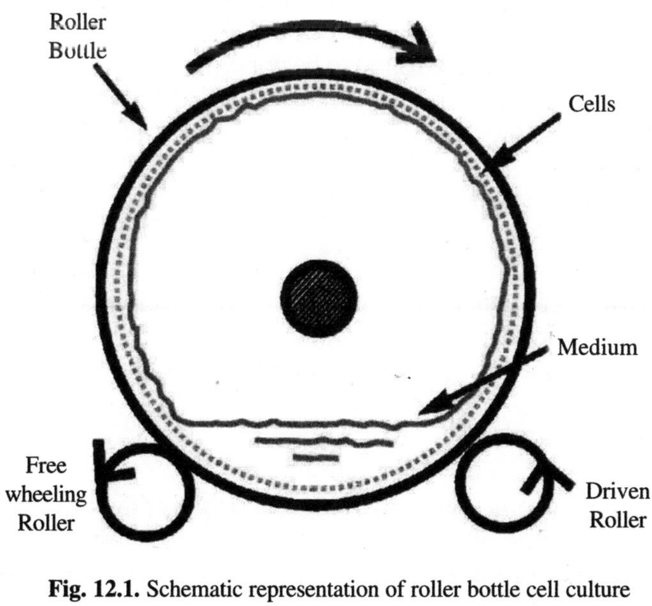

Fig. 12.1. Schematic representation of roller bottle cell culture

CHAPTER 13

CHARACTERIZATION AND PRESERVATION OF CELLS

13.1 Characterization of Cell Lines

After establishment of cell lines, these need to be characterized. The characterization includes confirmation of species or tissue of origin, and determination of differentiated status of cell within the lineage. The other important thing is to check for cross-contamination of cell types as for example fibroblasts tend to contaminate other cell types. The possibility of contamination is more in case the cell that has been cultured from wrong species or wrong tissues. Different studies indicates that more than 30% cell lines are found to be contaminated from another species or another cell lines from same species. Cell lines should, therefore, be periodically checked against this and monitored for instability and variation.

13.1.1 Species verification

When working with a specific cell line, one should know the species of origin of that cell line. Still there is need for verification of species as there are always chances of cross-contamination. The species of origin can be verified by methods such as cytogenetics, isoenzymology and immulogical tests.

13.1.2 Cytogenetic analysis

The analysis of chromosome is the most reliable method of distinguishing between species, as chromosome content is the most characteristic trait of a species. The preparation and study of karyotypes is part of cytogenetics. Karyotyping is a reliable method based on analysis of karyotype using both banding and non-banding pattern. Species identification rest on the ability to distinguish the chromosome banding pattern of different species.

13.1.3 Karyotyping techniques

Karyotypes describe the number of chromosomes, and what they look like under a light microscope. Attention is paid to their length, the position of the centromeres, any differences between the sex chromosomes, and any other physical characteristics.

Karyotype should be analyzed at specific stage of cell cycle as chromosome shape and size varies with the stage of nuclear division. Mainly metaphase stage is studied. It involves arresting of cells in metaphase using spindle fiber inhibitor such as colcemid/vinblastine.

G-banding based on giemsa staining is used to indentify the individual chromosome. Also C-banding which stains the centromere region and Q-banding based on quincrine mustard which the interband region can be used. Q-banding is particularly used for Y chromosome. The Y-chromosome is strongly fluorescent and very distinctive with Q-banding, while G-banding produces uniformly stained Y chromosome.

13.1.4 Characterization of cell type and stage of differentiation

Once the species of origin of cell line has been confirmed, the characterization of cell type and stage of differentiation can be easily confirmed by using functional assays such isoenzyme analysis, human lymphocyte antigen (HLA) typing, chromosomal analysis, karyotyping, morphology and STR analysis. Before we start looking at specific tests for determining tissue types, it is important to find out the major cell types, which exist *in vivo* and the processes by which different cell types arise.

13.1.5 Verifying origin of tissue

Cells are organized into tissues (groups of similarly specialized cells and associated extra cellular matrix) in the body which perform a specific function or group of functions. Each kind of tissue is composed of cells with a characteristic size, shape and arrangement. Several types of tissues unite to form an organ and organs are arranged into organ systems. Tissue is a cellular organizational level intermediate between cells and a complete organism. Hence, a tissue is an ensemble of cells, not necessarily identical, but from the same origin, that together carry out a specific function. Tissues can be grouped into four basic types: connective, muscle, nervous, and epithelial. Multiple tissue types comprise organs and body structures. While all animals can generally be considered to contain the four tissue types, the manifestation of these tissues can differ depending on the type of organism. Animal tissues may be classified as epithelial, connective, muscular, nervous and blood and lymph.

Connective tissue

Connective tissues are fibrous tissues which join together different tissues of the body, support the body and protects the underlying organs. They are made up of relatively few cells embedded in an extra cellular matrix. Connective tissue gives shape to organs and holds them in place. It is important to remember that there are many different types of connective tissue and including adipose tissue, cartilage and bone.

Muscle tissue

Muscular tissue is composed of cells specialized in contraction and form the active contractile tissue of the body. They contain contractile fibres called myofibrils which are composed of actin and myosin. Muscle tissue functions to produce force and cause motion, either locomotion or movement within internal organs. Muscle tissue is separated into three distinct categories: visceral or smooth muscle, which is found in the inner linings of organs; skeletal muscle, in which is found attached to bone providing for gross movement; and cardiac muscle which is found in the heart, allowing it to contract and pump blood throughout an organism.

Nervous tissue

Cells comprising the central nervous system and peripheral nervous system are classified as neural tissue. In the central nervous system, neural tissue forms the brain and spinal cord and, in the peripheral nervous system forms the cranial nerves and spinal nerves, inclusive of the motor neurons. The actual conducting cells of the nervous system are called neurons and they typically consist of a cell body, dendrites and axons. The connecting and supporting cells are called glial cells.

Epithelial tissue

The epithelial tissues are formed by cells that cover organ surfaces such as the surface of the skin, the airways, the reproductive tract, and the inner lining of the digestive tract. The cells comprising an epithelial layer are linked via semi-permeable tight junctions; hence, this tissue provides a barrier between the external environment and the organ it covers. In addition to this protective function, epithelial tissue may also be specialized to function in secretion and absorption.

Many epithelial cells are specialized to secrete a specific product, such as milk, mucus or sweat. There are different types of epithelium such as squamous, cuboidal and columnar. Epithelial tissue helps to protect organisms from microorganisms, injury, and fluid loss.

Blood and lymph tissue

Blood and lymph tissue includes all cells that are present in the peripheral blood and their precursor cells in the bone marrow and lymph nodes. The function of this tissue is to exchange oxygen and carbon dioxide with the external environment and for transport and immune response.

13.2 Cell Freezing and Recovery

The cells need to be preserved when they are not required as a sufficient amount of time, effort and money would be required to maintain the cells in culture when they are not required. So the cells are cryopreserved and revived back when needed.

13.2.1 Storage of cells in frozen state

Cryopreservation is a process where cells or tissues (primary samples) are preserved by cooling to low sub-zero temperatures, such as –196°C of liquid nitrogen in a state of suspended animation until they are needed. At these low temperatures, any biological activity, including the biochemical reactions that would lead to cell death, is effectively stopped. However, when cryoprotectant solutions are not used, the cells being preserved are often damaged due to freezing during the approach to low temperatures or warming to room temperature. Cryogenic storage at very low temperatures is presumed to provide an indefinite, if not near infinite, longevity to cells although the actual "shelf life" is rather difficult to prove. The main techniques to prevent cryopreservation damages are a well established combination of controlled rate and slow freezing and a newer flash-freezing process known as vitrification. New cell lines should be preserved in original state before they become genescent and loose their original functions. Great deal of caution should be taken while working with liquid nitrogen using eye protection and heavy gloves.

13.2.2 Cryoprotectant

Many of the damaging effects of cryopreservation like solution effects, extra-cellular ice formation, dehydration and intra-cellular ice formation can be reduced by cryoprotectants. The main function of the cryoprotectant is to replace the water in the cell and thus reducing the water content of the cells. Cells are suspended in a solution of cryoprotectant such as glycerol or dimethyl-sulfoxide (DMSO) in high concentration of serum. Then cells are first cooled at a defined rate in liquid nitrogen vapor and then transferred in liquid nitrogen container. DMSO is a small molecule, insoluble in lipid, hence enters cells by diffusion across the lipid bilayer of the cell membrane. Thus DMSO prevents the formation of harmful ice crystals. The high serum concentration contributes to cell integrity by maintaining the intra cellular protein concentration of cells. Glycerol also has similar effect.

13.2.3 Freezing mixture

The freezing mixture is a double concentrated (2x) solution containing 40% medium (with 10% FCS), 40% FCS and 20% DMSO (or glycerol). The mixture is made adding the components in the above order and then mixed gently by inversion. The freezing mixture should be filtered by a 2.0 μm sterile disposable filter and stored at –20°C. The freezing mixture is mixed with an equal volume of cell suspension in complete medium during cryopreservation.

13.2.4 Freezing down cells

Cells are cryopreserved only when they are in exponential phase because the confluent and overgrown cells would not recover well from cryopreservation. The cells should be counted and then centrifuged at 150–200 g at 4°C for 5 min. Discard the supernantent while not disturbing the cell pellet and then resuspend the pellet in the residual media by gentle tapping and adjusted to double the final concentration required with fresh, ice-cold media containing 10% FCS. Adherent cells should be 2×10^6/ml and suspension cells 1×10^7 cells after resuspension. Add equal volume of freezing mixture to the cell suspension placed on ice and mixed thoroughly. Aliquots 1 ml each of cell suspension into cold prepared cryotubes (with a sterile 1 ml plastic pipet or micro-pipet) maintaining the final cell concentration about 1×10^6 (adherent) and 5×10^6 (suspension) per vial. Don't over tighten the screw cap of cryotube as this may distort the gaskets and cause liquid nitrogen to enter the vials. A freezing plug allows cooling of the cells at a defined rate in the liquid nitrogen vapor over the liquid nitrogen within the storage tank. The cooling rate should be 1°C/min.

13.2.5 Revival of frozen cells

The cryovial should be collected from the liquid nitrogen storage tank with care and thawed quickly in water bath at 37°C. Then wipe the vial with 70% ethanol and taken into the hood. The media also should be pre-warmed to 37°C in a water bath. The contents of the vial should be transferred into a 15 ml sterile centrifuge tube containing 9 ml complete medium, the cap secured, and the contents mixed by inversion. The cells must be immediately removed from the freezing mixture in which they were stored by centrifugation at 150–200 g for 5 min. and then resuspended in 10 ml of complete medium.

A sample should be taken for counting and to check for viability before incubation in an appropriate flask at a sub-confluent concentration for 24 hr. The culture should be observed under the inverted microscope and counted again if necessary. Different Cell lines recover from liquid nitrogen with differing efficiencies. The cultures should be observed carefully for a number of days and sub-cultured when necessary. Some lines give a large percentage of dead cells after 24 hr and take 3–4 days before the culture looks healthy and confluent, while others recover almost immediately.

13.3 Transport of Cell Cultures and Cell Suspensions

Cells should be grown to near confluency in monolayer or to desire density in suspension before ready for transport. Both monolayer and suspension cultures can easily be shipped in 25 cm^2 tissue culture flasks. Replace the media with the fresh media and flask should be completely filled with medium to protect the cells from drying during transport. The cap should be tightened and sealed in a leak proof plastic bag or container designed to prevent spillage in the event of damaged flask. The container is placed in an insulated container to protect from extreme temperature during transport. A biohazard label is affixed to the outside of the package. Generally, cultures are transported by same day or overnight courier. Cells can also be shipped frozen. The vial containing frozen cells is removed from the liquid nitrogen freezer and placed immediately on dry ice in an insulated container to prevent thawing during transport.

13.4 Managing Cell Cultures

13.4.1 Obtaining the basic material: Importance of cell culture collections

The cell lines are stored in cell culture collections banks all around the world. Realizing the fundamental requirement for the application of sensitive and extensive quality control testing procedures, scientists have worked together to establish banks of tested cells. With the increased biotechnological use of animal cells, there has been a parallel increase in the number of culture collections and other resource centers *e.g.* American type culture collection (ATCC), European type culture collection (ECACC) etc.

13.4.2 Cell culture databases

With the availability of on-line data systems for the culture collections, the user has the privilege to access to the most up-to-date listing of material available. The

American Type Culture Collection (ATCC) database is available on-line at www.atcc.org, also The European Collection of Cell Cultures (ECACC) catalogue are available at www.ecacc.org.uk. Both ECACC and ATCC have recently introduced their main catalogues on CD-ROM, which provide for easier distribution and enable easy updates.

CABRI (Common Access Biotechnological Resources and Information)

CABRI stands for Common Access to Biotechnological Resources and Information an online service where users can search a number of European Biological Resource Centre catalogues. In identifying cell lines through CABRI only a single search is necessary without going through all the individual databases of the collections concerned. Currently ECACC, DSMZ, and IST are the cell line catalogues incorporated. The database is available on www.cabri.org. The service has been designed, implemented, and is maintained by a number of European Biological Resource and Information Centres, all of whom are committed to the highest quality standards. CABRI currently contains the catalogues of:

BCCM	Belgian Co-ordinated Collections of Micro-organisms Brussel, Belgium CABI Bioscience Egham, UK
CBS	Centraalbureauvoor Schimmelcultures Utrecht, The Netherlands Utrecht,
CIP	Institute Pasteur, Paris, France
DSMZ	Deutsche Sammlung von Mikroorganismen und Zellkulturen Braunschweig, Braunschweig, Germany
ECACC	Salisbury, UK
NCIMB	National Collections of Industrial, Marine and Food Bacteria Aberdeen, Scotland, UK
INRC	Genoa, Italy
NCCS	National Centre for Cell Science NCCS Complex, University of Pune Campus, Ganeshkhind, Pune, Maharashtra, India

NCCS (National Centre for Cell Science)

The National Centre for Cell Science (NCCS) formarly known as National Tissue Culture Facility in the year 1986 with a mandate of basic research, teaching and training, and as a national repository for cell lines/hybridomas etc.

During its transformation, the centre has undergone several changes to facilitate achievement of its goals. The centre has expanded its scientific charter to strengthen basic research in the areas of cancer biology, cell biology, immunology, genomics and proteomics.

It is very important to handle the cell lines with utmost care as a little negligence or contamination can destroy the entire cell line. A flow chart of the scheme which operates at the type culture collections with the essential points being:

- Cultures should be handled in Class II safety cabinets offering operator protection as the source of the cell lines may not be known.
- Cultures should be handled in a quarantine laboratory separate from the main tissue culture area.
- Initial characterization (*e.g.* species verification) and microbial quality control should be performed.
- After satisfying the minimum conditions the cultures may be transferred to the main tissue culture area for production of master and working cell cultures.

It is recommended that these procedures are followed irrespective of the source of the cell line, *i.e.* whether it is from a recognized culture collection/bank or from a known research laboratory.

CHAPTER 14

CELL LINES BANK

14.1 Introduction

In order to ensure consistency of the properties of a cell line, it is necessary to characterize and cryopreserve a reference stock of cells. Since some of the properties of a cell line are likely to change during an extended period of continuous passaging. Cell cultures are discarded usually after a defined passage interval and the cryopreserved stocks are used to reinitiate fresh cultures. A two tiered cell bank system is preferred to ensure continuity of supply of a cell line from a culture collection or for the manufacture of some biopharmaceutical product. A well characterized cell bank provides a consistent source of contaminant free cell line and may consists of a Master Cell Bank (MCB) and Working Cell Banks (WCB).

14.2 Strategies of Cell Line Banking

Production of cell banks

Characterization and safety testing of cell banks is an essential part of the safety concept for cell-derived biopharmaceutical products. The objective of this testing is to confirm identity, purity and the genetic stability of the cell line. The potential risk of introducing adventitious agents such as bacteria, mycoplasma, fungi or viruses to the biopharmaceutical must be eliminated or at least minimized. Therefore, testing is performed at the level of the master cell bank, the working cell bank, and the end-of-production cells or post-production cells.The test package for the characterization of a cell line is species-specific. The source of the cell line and its documented history and development as well as the biological properties of the cells also determine which testing should be performed.

When a culture is received in the laboratory a strict accessioning scheme should be followed. The cultures received should be propagated as per the conditions recommended by the supplier. Further this should be subjected to detailed microbial quality control and characterization. When all tests provide satisfactory results then a seed stock can be made. All steps should be carried out in the quarantine laboratory designated for the purpose. Also master cell bank should carry out major authentication aspects such as isoenzyme analysis, karyology, and DNA finger printing analysis.

The objective of cell bank characterization is to ensure that the cells in the cell bank meet the following criteria:

- That the cells are viable, free from contaminants and are capable of further propagation
- The cells must have the desired characteristic(s) for which they were selected and can be used for wider range of applications.
- Can be used as a reference.

Baseline data on cell line characterization is desired as such a reference database is essential because the cell bank curator will have the unenviable task of ensuring that their materials have been appropriately stored for long duration.

Some points to be considered when working with cell banking:

1. Cells should be healthy with a viability of >90% and no signs of microbial contamination.
2. Cells should be in log phase of growth (using pre-confluent cultures *i.e.* cultures that are below their maximum cell density).
3. Use of cryoprotectant such as dimethyl sulphoxide or glycerol to help protect the cells from rupture by the formation of ice crystals. The most commonly used cryoprotectant is DMSO at a final concentration of 10%, an alternative cryoprotectant such as glycerol can also be used.

Ultra-low temperature storage of cell lines

Following controlled rate freezing in the presence of cryoprotectants, cell lines can be cryopreserved in a suspended state for indefinite periods provided a temperature of less than –135°C is maintained. Such ultra-low temperatures can only be attained by ultra-low freezers or more usually by immersion in liquid or vapor phase nitrogen. The advantages and disadvantages can be summarized as follows (Table 14.1).

Table 14.1 Comparison of Ultra-low Temperature Storage Methods for Cell Lines

Method	*Advantages*	*Disadvantages*
Electric (–135 °C) Freezer	• Ease of maintenance • Steady temperature • Low running costs	• Requires liqid nitrogen back-up • Mechanically complex • Storage temperatures high relative to liquid nitrogen
Liquid Phase Nitrogen	• Steady ultra-low (–196 °C) temperature • Simple	• Requires regular supply of liquid nitrogen • High running costs • Risk of cross-contamination via the liquid nitrogen
Vapor Phase Nitrogen	• No risk of cross-contamination from liquid nitrogen • Low temperatures achieved • Simplicity and reliability	• Requires regular supply of liquid nitrogen • High running costs • Temperature fluctuations between –135°C and –190°C

Storage in liquid phase nitrogen allows the lowest possible storage temperature to be maintained with absolute consistency, but requires the use of large volumes (depth) of liquid nitrogen. For this reason and also risk of contamination by virus pathogen, ultra-low temperature storage is most commonly preferred in vapor phase nitrogen. For vapor phase nitrogen storage, the ampoules are positioned above a shallow reservoir of liquid nitrogen, the depth of which has to be carefully maintained. A vertical temperature gradient will exist through the vapor phase, the extremes of which will depend on the liquid levels maintained, the design of the vessel, and the frequency with which it is opened.

Liquid nitrogen storage vessels with alarms that at least warn of low liquid nitrogen levels is preferred. This is particularly true of vapor phase storage systems. The bulk liquid nitrogen storage vessel should not be allowed to become less than half full before it is resupplied.

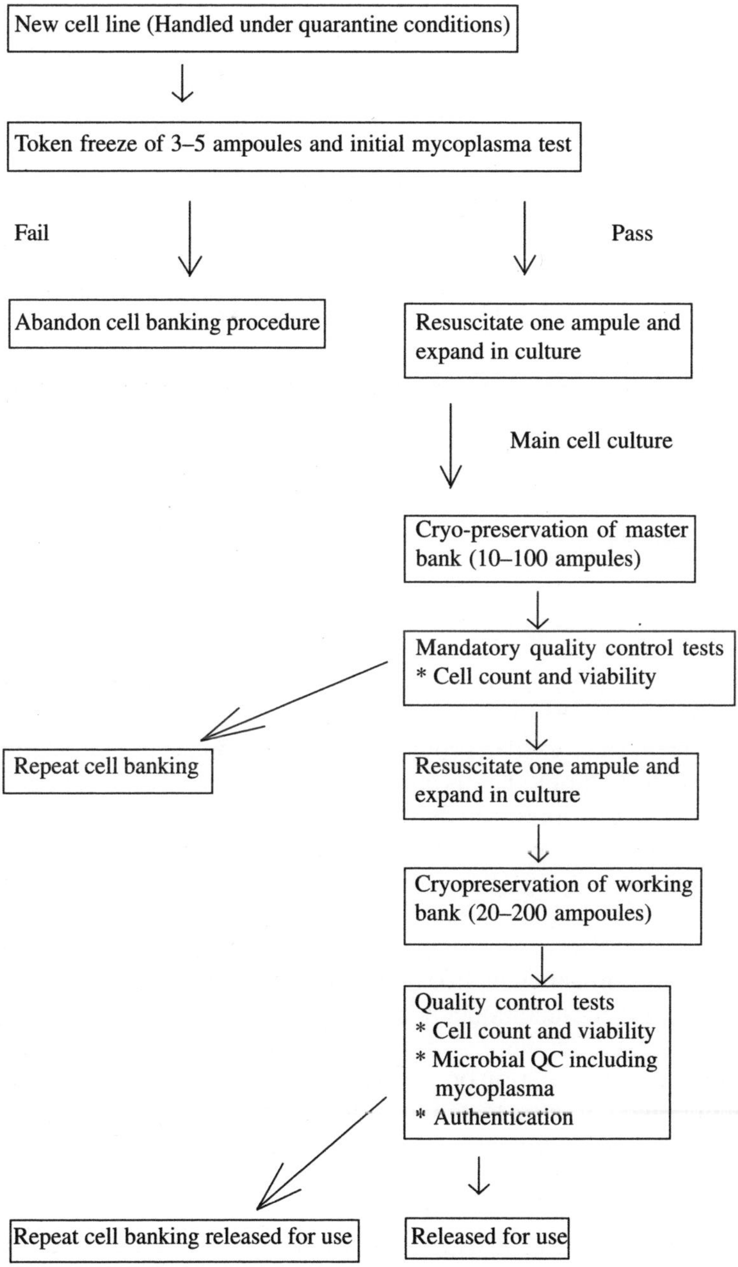

Basic strategy for cell banking

CHAPTER 15

FLOW CYTOMETRY

15.1 Introduction

Flow cytometry was developed in 1970's which was employed for studying single cell quantitation, was rapidly become an essential instrument for research in the field of biological sciences. It is an accurate and rapid method for analyzing individual cells in a suspension. In this technique several parameters can be detected simultaneously and analyzed to provide information on the relationship between them. Fluorescently labeled cells can give quantitative data on specific target molecules and their distribution in the cell population. Subpopulations of cells with specific attributes can be discriminated and physically sorted for further development.

Generally the cells in culture or in fermentors are studied by using microscopic and biochemical approaches such as metabolic or enzymatic assays. But there are severe limitations to the techniques such as the microscopic methods are slow and provide only the static, qualitative information. However the biochemical methods do not provide any information on the heterogeneity of the cell population. With the advent of flowcytometric techniques, abilities to identify and characterize the morphological and biochemical heterogeneity of cell populations have advanced, permitting a rapid and sensitive cell analysis.

Flow cytometry (FCM) is widely used in health research and treatment for a variety of tasks, such as providing the counts of specific cell types which are needed to monitor the course of treatment and diagnosis in many diseases.

The technique of analyzing individual cells in a fluidic channel was first described by Wallace Coulter in the 1950s, and applied to automated blood cell counting.

The present "state-of-the" flow cytometers are capable of analyzing 10 to 20 parameters per cell at rates up to 100000 cells per second. Various immunofluorescent dyes or antibodies can be attached to the antigen or protein of interest. The suspension of cells or particles is aspirated into a flow cell where, surrounded by a narrow fluid stream, they pass one at a time through a focused laser beam. The light is either scattered or absorbed when it strikes a cell. Absorbed light of the appropriate wavelength may be reemitted as fluorescence if the cell contains a naturally fluorescent substance or one or more fluorochromes-labeled antibodies are attached to surface. Light scatter is dependent on the internal structure of the cell and its size and shape. Fluorescent substances absorb light of an appropriate wavelength and reemit light of a different wavelength.

15.2 Principle of Flow Cytometry

Flow cytometry has the ability to measure the properties of individual particles. The samples when injected into a flow cytometer (Fig. 15.1), the particles are distributed randomly in 3D space. In order to interrogate single particle the sample must be having ordered stream of single particles. The fluidics system includes a central channel or core for sample injection which is enclosed in an outer sheath that contains fast flowing fluid. When the fluid in the sheath moves it creates a massive drag effect on the central chamber which alters the velocity of the central fluid by which the flow front becomes parabolic with greatest velocity at it center and zero velocity at the wall. This effect is called hydrodynamic focusing.

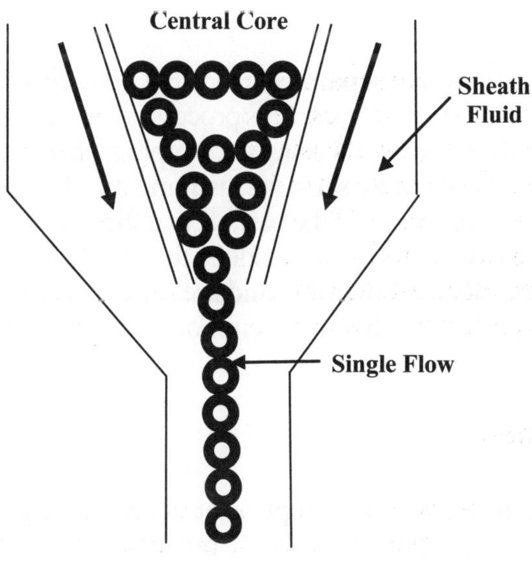

Fig. 15.1. Basic design of flow cell

The flow characteristics of the center fluid can be calculated by Reynolds number $Re = pVD/\mu$ where D is tube diameter, V the mean velocity of fluid, p is density of fluid, and μ the viscosity of fluid. When $Re < 2300$, flow is always laminar while $Re > 2300$ flow can be turbulent which accelerates diffusion.

After hydrodynamic focusing, each particle passes through one or more beams of light scattering or florescence emission which provides information about the particle's properties. Laser and lamp are the most commonly used light sources. Fluorescence measurement taken at different wavelengths can provide quantitative and qualitative data about fluorochromes labeled cell surface receptors or some of the intracellular molecules *e.g.* DNA, cytokines etc. Detectors are either silicon photodiodes or photo multiplier tubes (PMT).

15.2.1 Signal processing

A small current is generated when the light hits a photo detector. The associated voltage has amplitude proportional to the total number of light photons received by the detectors. This voltage is then amplified by a series of linear or logarithmic amplifiers. For DNA analysis linear scaling is preferable. The measurements from each detector are called parameter *e.g.* forward scatter, side scatter and fluorescence. The data acquired in each parameter is called events and is proportional to number of cells having the physical characteristics or marker of interest.

15.2.2 Cell sorting

In flow cytometry the cells are separated according to their subtype or epitope expression in case of biological studies. This process is called cell sorting or FACS analysis. After hydro dynamically focusing the sample each particle is probed with a beam of light. The sort criteria are set in the instrument with which the scatter and fluorescence signal are compared. If the particle matches the selection criteria the fluid stream is charged as it exits in the nozzle of the fluidics system. The droplets pass through a strong electrostatic field and are reflected left or right based on their charge. The speed of flow sorting depends on particle size and rate of droplet formation.

15.2.3 Fluorochromes

Fluorochromes are dyes, which accept light energy at a given wavelength and re-emitted at a longer wavelength. These processes are called excitation and emission. Emission is a very rapid process which occurs in nanoseconds, in the case of fluorochromes it is called fluorescence. There is a range of fluorescent molecules with a potential application in flow cytometry which can be categorized as: 1) Single dye *e.g.* FITC, and 2) Tandem dyes.

Firstly the dye molecules are excited and when it reaches the maximal singlet state where all its energy is transferred to the second dye molecules. This activates the second fluorochrome molecules which produces the fluorescence emission; this process is called FRET (Fluorescence resonance energy transfer). The commonly used dyes are listed below (Table 15.1):

Table 15.1 Some Fluorescent Dyes used in Flow Cytometry

Dye	*Laser excitation line (nm)*	*Maximal absorbance (nm)*	*Maximal emission (nm)*	*Fluorescence color*
Alexa Fluor 405	405,407	401	421	Blue
Alexa Fluor 430	405,407	433	541	Yellow
Alexa Fluor 488	488	495	519	Green
Alexa Fluor 633	633,635,647	632	647	Violet
Alexa Fluor 647	633,635,647	650	665	Violet
Alexa Fluor 660	633,635,647	663	690	Violet
Alexa Fluor 680	633,635,647	679	702	Violet
Alexa Fluor 700	633,635,647	702	723	Infrared
APC	633,635,647	650	661	Violet
FITC	488	490	525	Green
Pacific Blue	405,407	410	455	Blue
Per CP	488	490	675	Red
PE	488	490, 565	578	Yellow

Abbreviations: APC –allophycocyanin; FITC- fluorescein isothiocyanate; PE- phycoerythrin; PerCP-Peridinin-chlorophyll-protein complex

15.3 Data Analysis

In flow cytometry (FCM), intact cells and their constituent components are tagged with fluorescently conjugated monoclonal antibodies and are stained with fluorescent reagents and then analyzed individually by a flow cytometer. In the instrument, hydrodynamic forces align the cells and the fluorescent molecules on each cell are excited by passing through the laser light at speeds exceeding 70000 cells per second. Each cell passing through the beam also scatters light providing an indication of cell shape and size. A flow cytometer is capable of measuring up to 20 cell characteristics, for up to millions of individual cells per sample aliquot. This technology can be used to examine many cellular parameters on live or fixed cells, including surface, cytoplasmic, and nuclear proteins, DNA, RNA, reactive-oxygen species, intracellular pH, and calcium flux. Measurement of the expression of cellular-activation markers, intracellular cytokines, immunological signaling, and cytoplasmic and nuclear cell cycle and transcription factors can also be readily performed.

Typical FCM data analysis involves: (i) **Gating** (*i.e.* identification of homogenous cell populations that share a particular function); (ii) **Interpretation** *i.e.* finding correlations between some characteristics of the identified cell populations (*e.g.* percentages of cells in a cell population, median fluorescent intensity of a cell population for different markers) and clinical outcomes.

FCM data file is analyzed through a series of analysis components (shown in Fig. 15.2). This framework is constructed to report details of FCM data analysis studies in a systematic way to facilitate reporting and review process.

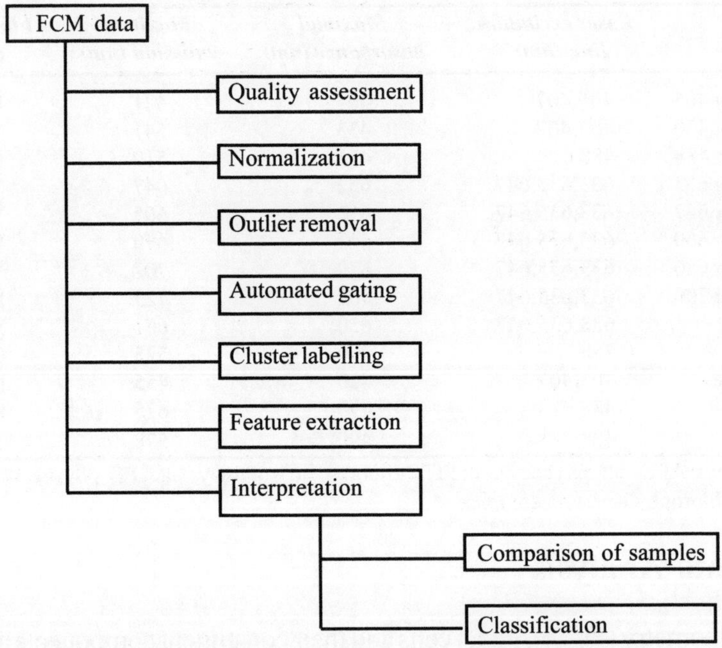

Fig. 15.2 Flow chart of analysis of FCM data file

15.3.1 Quality assessment

Artifacts from sample preparation, handling, variations in instrument parameters, or other factors may confound experimental measurements and lead to erroneous conclusions. The aim of data quality assessment could include detecting whether intersample variability measurements of samples are not likely to be biologically motivated. Such samples should be identified, investigated, and potentially removed before any further analysis.

15.3.2 Normalization

Like all other high-through put data sources, there is a substantial need for normalization steps to remove non-biological variations so that the analysis can focus on the important and relevant biological variations between samples. Instrument variability (*e.g.* changes in laser power), experimental protocol changes (*e.g.* changes in voltage setting of the instrument), and reagent changes (*e.g.* using antibodies from different sources) are examples of non-biological factors that can introduce variability in the data and shift the location of cell populations.

15.3.3 Outlier removal

Outliers refer to observations (events in the FCM data) that deviate to such a large extent from others so as to arouse suspicion that they do not belong to the same group of observations of interest. Cell debris, dead cells, and doublets (multiple events at the same time) often contaminate FCM data and give rise to outliers. Statistics derived from data sets that include outliers may be misleading. Therefore, it is crucial to identify outliers and account for their prevalence so as to minimize their effect on subsequent analysis.

15.3.4 Automated gating

Automated identification of homogenous cell populations that share a particular function is referred to as automated gating. The main purpose of automated gating is having an objective and systematic approach for classifying cells.

Automated gating can be used to (i) identify known cell populations and (ii) discover new subpopulations of cells that might not be easily detected via standard manual gating methods. For example, cell populations may be missed due to limitations of two dimensional manual gating.

15.3.5 Cluster labeling

Comparison of FCM samples is only possible if the same cell populations of different samples are compared against each other. For example, lymphocyte cells of two different samples can be compared against each other. Cluster labeling is referred to the procedure of finding similar cell populations between samples after automated gating.

15.3.6 Feature extraction

This step involves computing measurable heuristic properties (also referred to as features) of the identified gates for further analysis. Percentages of cells with respect to the total number of cells, median and standard deviations of fluorescent intensities of different markers for the events within each gate (or gates of interest) are examples of features that can be computed for the next step.

Interpretation of gating results is highly dependent on what the objective of the study is. Usually, there are two major objectives in an FCM-based study:

(a) Statistical comparison of samples, where the samples are compared to see if they share similar characteristics;
(b) Classification, where the samples are labeled to predefined classes such as healthy versus patient or patients with short survival versus affected.

15.4 Preparation of Sample

Single cell suspension of density 110–120 cells per ml should be made for sample analysis. As the higher concentration of cells can narrow bores of the flowcytometer and its tubing can be blocked up. Also the cell concentration influences the rate of flow sorting. The higher sort speed may decrease purity of the sample. The samples can be non-adherent cells from culture, water born microorganism, bacteria, yeast or some solid tissue *e.g.* liver or tumors and even whole blood can be easily used as a sample for the analysis. In case of cells from suspension cultures, PBS (phosphate buffered saline) is commonly used as suspension buffer and in case of solid material, the sample needs to be disaggregated as to produce single cells either mechanically or enzymatically. In mechanical disaggregation the suspension of chopped tissue is passed through of fine gauges needle for several time followed by grinding and sonication if required in some cases. This process is used in case of loosely bound structures such as adherent cells from culture, bone marrow and lymphoid tissue. The enzymatic disaggregation method is used to disrupt protein-protein interaction and some extra cellular matrix and associated bio-molecules that help in cell intactness. The enzyme activity depends on some physiological factor such as pH, temperature and the various co-factors associated with the enzyme.

In case of studying intracellular components, the plasma membrane of the cell need to be permeablized to allow dye molecules or EDTA (ethylene diaminetetraacetic acid), EGTA (ethyleneglycol–bis (2-aminoethylether) N,N,N',N' tetra acetic acid antibody molecules through while retaining the cell integrity. Method to be employed for sample preparation depends on the starting material and the nature of the epitope.

15.4.1 Preparation of cells

(a) Cells stored in liquid nitrogen

Carefully remove cells from liquid nitrogen storage and thaw rapidly using PBS/BSA buffer (phosphate buffered saline pH 7.4 and 1% BSA) and place into a 15 ml conical centrifuge tube. Centrifuge at 400 g for 5 minutes and discard supernatant and resuspend pellet in an appropriate amount of PBS/BSA buffer.

(b) Tissue culture cell lines in suspension

Decant cells from tissue culture flask into 15 ml conical centrifuge tube and centrifuge at 400 g for 5 minutes. Then discard supernatant and resuspend pellet in 10 ml of PBS/BSA. Again centrifuge at 400 g for 5 minutes and finally discard supernatant and resuspend pellet in PBS/BSA.

(c) Adherent tissue culture cell lines

Gently scrape the harvested cells using 2 ml of PBS/BSA buffer and transfer cells to a 15 ml conical tube and add buffer to make the final volume to 10 ml. Then centrifuge at 400 g for 5 minutes and discard supernatant and resuspend pellet in fresh PBS/BSA (10 ml). Again centrifuge for 5 minutes at 400 g. Finally discard supernatant and resuspend pellet in PBS/BSA buffer.

(d) Preparing cells from solid/lymphoid tissues

Perfuse the tissue gently using a syringe and needle in 15 ml of PBS/BSA and transfer the cell suspension from the petri-plate into a 15 ml conical centrifuge tube. Centrifuge at 400 g for 5 minutes and resuspend the pellet in PBS/BSA. Then add 10 ml of ammonium chloride lysis buffer. Mix and incubate for 2 minutes. Again centrifuge at 400 g for 5 minutes then add 10 ml of PBS/BSA and mix properly. Again centrifuge at 400 g for 5 minutes resuspend the pellet to make the final volume 10 ml with PBS/BSA. Use hemocytometer for cell counting and adjust the cell suspension for cell count of $0.7–1.2 \times 10^7$cells/ml.

15.5 Applications of Flow Cytometry

(i) Cell biology
(ii) Clinical medicine
(iii) Plant and animal sciences
(iv) Pharmaceuticals
(v) Analysis of DNA Ploidy, cell cycle and cell death
(iii) In organ transplantation and hematopoietic cell therapy
(iv) Applications in Microbiology
(v) Determination of cell number and viability
(vi) Monitoring of cell line stability
(vii) Reproductive biology–Sperm sexing
(viii) Analysis of intracellular cytokines

REFERENCE

Abercrombie, M. and Heaysman, J. E. (1953). Observations on the social behavior of cells in tissue culture. Exp.Cell Res. **5**: 111–131.

Arnold, J. (1887). Ueber Theilungsvorgange an der Wanderzellen; ihre progressiven und regressiven Metamorphosen. Arch Mikrosk ant. **30**: 205–310.

Burrows, M. T. (1910). 'The cultivation of tissues of the chick embryo outside the body'. *J. Am er. med. Assn.* **55**: 2057–2058.

Carrel, A. (1912). On the permanent life of tissues outside the organism. *J. Exp. Med.* **15**: 516–528.

Carrel, A. (1913). "Artificial activation of the growth *in vitro* of connective tissue". *J. of Exp. Med.* **17**: 14.

Carrel, A. and Burrow, M.T. (1911). An addition to the technique of the cultivation of tissues *in vitro. J. of Exp. Med.* **14**: 244–247.

Dulbecco, R. (1952). Production of plaques in monolayers of tissue cultures by single particles of an animal virus. *Proc. Natl. Acad. Sci. USA*. **38**: 747.

Dulbecco, R. and Vogt, M. (1954). Plaque formation and isolation of pure cell lines with poliomyelitis viruses. *J.Exp.Med.* **199**: 167–182.

Eagle, H. (1955). The specific amino acid requirements of mammalian cells (stain L) in tissue culture. *J. Biol. Chem.* **214**: 839.

Eagle, H. (1959). Amino acid metabolism in mammalian cell cultures. *Science* **130**: 432–437.

Earle, W.R., Schilling, E.L., Bryant, J.C. and Evans, V.J. (1954). *J. Natl. Cancer Inst.* **14**: 1159.

Earle, W.R., Schilling, E.L., Stark, T.H., Straus, N.P., Brown, M.F., and Shelton, E. (1943). Production of malignancy *in vitro*; IV: The mouse fibroblast cultures and changes seen in the living cells. *J. Natl. Cancer Inst.* **4**: 1656–212.

Ebeling, A. (1913). The permanent life of connecive tissue outside of the organism, ibid. **17**: 273–285.

Fell, H.B. and Robison, R. (1929). The growth development and phosphatase activity of embryonic avian femora and limb buds cultivated *in vitro. Biochem. J.* **23**: 767–784.

Ham, R. (1965). Clonal growth of mammalian cells in a chemically defined synthetic medium. *Proc. Natl. Acad. Sci. USA*. **53**:288–293.

Ham, R.G. (1984). Growth of human fibroblasts in serum free media. In Barnes, D.W., Sirbasku, D.A. and Sato, G.H. (eds). *Cell culture methods for molecular and cell biology*. **Vol. 3** New York, Alan R. Liss, pp. 249–264.

Hanks, J.H. and Wallace, R.E. (1949). Relation of oxygen and temperature in the preservation of tissues by refrigeration. *Proc. Exp. Biol. Med.* **71**: 196.

Harris, M. (1959). Essential growth factor in serum dialyate for chick skeletal muscle fibroblasts. *Proc. Soc. Exp. Biol. Med.* **102**: 468.

Hayflick, L. and Moorhead, P.S. (1961). The serial cultivation of human diploid cell strains. *Exp. Cell Res.* **25**: 585–621.

John R.W. Masters (2000). Animal Cell Culture: A Practical Approach. Pub. Oxford University Press, USA.

Junggren, L.C.A. (1898). Von der Fahigkeit des Hautepithels, ausserhalb des Organismus sein Leben zu behalten, mit Berucksichtigung der Transplantation. Dtsch Z Chir. **47**: 608–615.

Kohler, G. and Milstein, C. (1975). Continuous culture of fused cells secreting antibody of predefined specificity. *Nature (London)* **256**: 495–496.

Kurotchkin, T.J. (1939). Proc. Soc. Exp. Bio. and Med. **41**: 407.

Lews, and Lewis. (1911). Grwoth of embryonic chick tissues in artificial media, agar, and bouillon. Johns Hopkins Hosp. Bull. **22**: 126.

McAteer , J.A. and Davis J. M. (1994). Basic cell culture technique and the maintenance of cell lines. In Basic Cell Culture; A practical approach (ed.) Davis J.M. Oxford University Press Inc., NY.

Moore, G.E., Gerner, R.E. and Franklin, H.A. (1967). Culture of normal human leukocytes. *J. Anim. Med. Assoc.* **199**: 519–524.

Morgan, J.G., Morton, H.J. and Parker, R.C. (1950). Nutrition of animal cells in tissue culture; I: Initial studies on a synthetic medium. *Proc. Soc. Exp. Biol. Med.* **73**: 1.

Moscona, A. (1952). Cell suspensions from organrudiments of chick embryos. Exp. Cell Res. **3**: 535.

Parker, R.C., Castor, L.N. and McCulloch, E.A. (1957). Altered cell strains in continuous culture. Special publications. *NY Acad. Sci.* **5**: 303–313.

Puck, T.T. and Marcus, P.I. (1955). A rapid method for viable cell titration and clone production with HeLa cells in tissue culture. The use of X-irradiated cells to supply conditioning factors. *Proc. Natl. Acad. Sci. USA*. **41**: 432–437.

Reitzer, L.J., Wice, B.M. and Kennel, D. (1979). Evidence that glutamine, not sugar, is the major energy source for cultured HeLa cells. *J. Biol. Chem.* **254**: 2669–2677.

Roax,W. (1885). Beitrage zur Entwickelungsmechanik des Embryo, Zeitschr. Biol. **21**: 411–526.

Rous, P. and Jones, F.A.A. (1916). A method of obtaining suspensions of living cells from the fixed tissues, and for the plating out of individual cells. *J. Exp. Med.* **23**: 555.

Tyrode, M.V. (1910). Arch. Int. Pharmacodyn. *Ther.* **20**: 205.

Willmer, E.N. and Kendal, L.P. (1932). *J. Exp. Biol.* **9**: 149.

Willmer, E. N. (1945). Growth and Form in Tissue Cultures. In 'Growth and Form Essays presented to D'Arcy Thompson', ed. W. Le Gros Clark and P. B. Medawar. Oxford (University Press). Tissue Culture', 2nd edition. London (Methuen).

Wilmut, A. E., Schnieke, J. McWhir, Kind A. J. and Campbell, K. H. S. (1997).Viable offspring derived from fetal and adult mammalian cells Nature **385**: 810–813.

SUGGESTED READING/REFERENCE

Adams R.L.P. (1983). Cell Culture for biochemists, Elsevier, New York.

Blaschko, H, Bonney, R. (1962). Proc. R. Soc. Lond. Ser. B. **156**: 268–279.

Bodziak, M.L., Grefrath, P.L., Price, P.J. and Jayme, D.W. (1985). In vitro, 21: 47A.

Cron, M.J., Godfrey, J. C., Hooper, I. R., Keil, I.G., Nettleton, D. E., Price, K. E. and H. Schmitz, Prog. (1970). Antimicrobial Anticancer Chemotherapy IL 1069.

Davis, J.M. (1994). Basic Cell Culture, 2nd Edn. Pub. Oxford University Press, NY.

Dawson, J.R., Adams, D.J. and Wolf, C.R. (1985). Induction of drug metabolizing enzymes in human liver cell line Hep G2, FEBS letters Research letters, **183** (2) 219–222.

Demain, A. L., Walton, R. B., Newkirk, J. F. and Miller, I. M. (1963). Nature, London, **199**: 909.

Docherty, R.J., Hamilton, A.O., Pattison, N., Mounsey, G., Foster, R., Hodgson, P., and Benny, G. (1994). FOCUS ®, **16**: 14–17.

Douglas and Dell' Orco (1979). Physical Aspects of a Tissue Culture Laboratory, in Methods in enzymology vol LVIII (William B. Jakoby, Ira H. Pastan., eds), Academic Press, New York pp 3–18.

Eagle, (1959). Science **130**: 432.

Evans, V. S., Bryant, J. C., Fioramonti, M. C., McQuilkin, W. T., Sanford, K. K. and Earle, W. R. (1956). Studies of nutrient media for tissue cells in vitro. I. A protein free chemically defined medium for cultivation of strain Leydig cells, Cancer Res.**16**: 77–86.

Freshney, R. I. (1986). Culture of Animal Cells: A manual of basic technique, 1st Edn. Pub. IRL Pres Ltd., England.

Freshney, R. I. (2005). Culture of Animal Cells: A manual of basic technique, 5th Edn. Pub. John Wiley and Sons, Inc, NJ.

Gilbert S.F. and Migeon, B.R. (1977). *J. Cell Physiology.* **92**: 161–167.

Gilbert, S.F. and Migeon, B.R. (1975). Cell. **5**: 11–17.

Ham, R. G. (1963). An improved nutrient solution for diploid Chinese hamster and human cell lines, Exp. Cell Res. **29**: 515–526.

Hodgson, J. (1991). Biotechnology, **9**: 1320–1324.

Holmes, R., Helms, J., and Mercer, G.. (1969). *J. Cell Biol.* **42**: 262–271.

Honn, K.V., Singley, J.A. and Chavin, W. (1975). Proc. Soc. Exp. Biol. Med., **149**: 344–347.

Hou, C. T. and Penman, D. (1970). *J. BioL Chem.* **245**: 1289.

Hutchings, S. E. and Sato, G. H. (1978). *Proc. Natl., Acad. Sci.* USA **75**: 901.

Jakoby, W.B. and Pastan I.H. (1979). Methods in Enzymology, Vol.LVIII - Cell Culture. Pub. Academic Press, New York.

Jayme, D.W. and Smith, S.R. (2000). Cytotechnology. **33**: 27–36.

Jennie, P. M. and David, B. (2006). Animal Cell Culture Methods.Pub.: Elsevier Science and Technology Books.

Lambert, K.J. and Birch, J.R. (1985). Animal Cell Biotechnology, **1**: 85.

Lillian E. Baker and Alexis Carrel. (1928). Nitrogen metabolism of normal and sarcomatous fibroblasts in pure cultures. *J. Exp. Med.* **48** (4): 533–547.

Michael Butler (2004). Animal Cell Culture and Technology, 2nd Edn.Pub. Garland Science/ BIOS Scientific Publishers, UK and New York.

Morgan, S.J. and Darling, D.C. (1993). Animal cell culture: Pub. BIOS Scientific Publishers, Oxford.

Nigel Jenkins (1999). Animal Cell Biotechnology: Methods and Protocols. Pub: Humana Press, NJ.

Perlman, (1979). Use of Antibiotics in Cell Culture in Methods in enzymology, vol LVIII (William B. Jakoby, Ira H. Pastan.,eds), Academic, New York press pp 110–116.

Perlman, D. (1964). Antimicrobial Agents and Chemotherapy Science. 144 (3616) 322–323.

Poiley, J.A., Schumann, R. F., and Pienta, R. J. (1978). *In vitro*, **14**: 405–412.

Price, P.J. and Gregory, E.A. (1982). *In vitro* **18**: 576–584.

Rossi, C.R., Keisel, G.K. (1980). *Am. J. Vet. Res.* **41**: 557–560.

Waymouth, C. (1959). Rapid proliferation of sub-lines of NCTC Clone 929 (Strain L) mouse cells in a simple chemically defined medium (MB 752/1) *J. Natl. Cancer Inst.* **22**: 1003–1015.

Yeoman, D. (1994). Animal Cell Technology, Basic and Applied Aspects (Kobayashi, T. et al. (eds.), **6**: 359–363.

GLOSSARY

Term		*Description*
Aliquot	:	A small sample
Animal cell technology	:	That part of biotechnology which uses animal cells, propagated *in vitro*, for the manufacture of bio-products and as tools in the discovery and testing of medicines.
Antigen	:	A molecule capable of inducing an immune response
Antigenic determinant	:	The structural component of an antigen molecule that is responsible for specific interaction with antibody molecules elicited by the same or a related antigen.
Ascites	:	The accumulation of serous fluid in the abdominal cavity of a mammal.
Aseptic	:	Without microbial contamination
Attenuation	:	Decreasing the virulence of a pathogenic organism.
Autologous	:	The same as, *i.e.* cells from the same persons or strain of animals
Bioreactor	:	A closed vessel for the cultivation of cells under controlled conditions.
Buffer	:	A liquid which resists changes in pH.
Cell bank	:	A uniform pool of cells, distributed into vials and preserved, typically by freezing in liquid nitrogen. A vial is thawed to start each new production cycle.
Cell culture	:	To grow cells *in vitro*
Cell cycle	:	The process of cell division, comprising four parts (G_1, S, G_2, M) which describe DNA replication and cell division
Cell line	:	An established in vitro cell type (usually immortal or transformed).
Cloning	:	Cell culture: selection of a single daughter cell from which to establish a culture of identical cells: molecular biology, manipulation of DNA replicas of DNA or RNA found in a tissue, organ or cell type.
Confluence	:	The measure of density of a cell culture
Continuous cell line	:	A cell line which has acquired the characteristic of being 'immortal', *i.e.* capable of growing indefinitely in culture.
Crisis (cell)	:	The senescence or death of a cell culture (sometimes followed by an outgrowth of immortal cells).

Table (*Contd...*)

Table *(Contd...)*

Term		*Description*
Cryopreservation	:	To preserve by freezing at low temperature
Cytotoxicity	:	To kill a cell (by another cell or drug)
Differenciation	:	A change in a cell causing it to lose or gain cell functions (usually non-reversible).
Epitope	:	An antigenic determinant of known structure.
Expression	:	The formation of proteins and peptides in a cell or on a cell surface by translation of the genetic code.
Fusogen	:	A substance which will cause membranes of adjacent cells to merge
Gene therapy	:	Modification of genetic information of somatic cells of the patient with the purpose to replace or supplement defective or missing parts of cellular DNA.
Gene	:	The biological unit of heredity, containing information for structure and function. A gene is composed of a specific sequence of nucleic acids.
Genetic engineering	:	A term which comprises methods for (1) the isolation and recombination of genetic material (including recombination of DNA from different species), (2) the insertion of recombinant DNA into an organism, either the original one or one in which this DNA does not naturally occur (host cell) and (3) the replication and expression of this DNA in its new environment.
Genome	:	The DNA content of a cell
Genotype	:	The gene content of a cell
Host system	:	A cell type which is used as the carrier for DNA from another source.
Humanising	:	Adaptation of non-human genes to the human genetic code, to enable animal cells or bacteria to express human-like molecules.
Hybrid	:	The result of two or more cells fusing to form a single entity
Hybridoma	:	A cell line resulting from the fusion of an antibody secreting blood cell (lymphocyte) and an immortal myeloma cell line. Produces a 'monoclonal' antibody.
Immortalization	:	Changing a primary cell type with a limited life span *in vitro*, to a cell type with unlimited capacity to proliferate
Immunization	:	To create an immune reaction by introducing a known antigen
Immunoglobulin	:	A protein of animal origin endowed with known antibody activity, synthesized by lymphocytes and plasma cells. They function as specific antibodies and are responsible for the humoral (blood-related) aspects of immunity.
Immunology	:	That branch of medical science concerned with the response of the organism to antigenic challenge, the recognition of self from non-self, and all the biological, serological and physical-chemical aspects of immune phenomena.

Table *(Contd...)*

Table (*Contd...*)

Term		***Description***
In vitro	:	Cell growth outside the body ('in glass')
In vivo	:	Cell growth within the body ('in life')
Karyotype	:	The chromosome complement of an individual cell.
Ligation	:	Use of the enzyme ligase to join separate DNA molecules to form a single DNA molecule.
Lymhokines	:	A 'family' of proteins produced by lymphocytes (blood cells) which have a wide range of regulatory functions within the body.
Magic bullet	:	The concept describing the linking of a specific antibody (often a monoclonal antibody) to a toxin or drug with the purpose of bringing the toxin or drug specifically to the target cell or tissue.
Medium	:	A buffered selection of components to support the growth of cells outside their normal environment
Medium	:	fluid composed of all nutrients necessary for cell cultivation.
Mitogen	:	Stimulator of cell proliferation
Monoclonal antibodies	:	Antibodies derived from a single cell type producing a single antibody type
Monolayer	:	Growth of adherent cells on substratum
Nutrients	:	Substances necessary for growth, normal functioning and maintaining life. Most often amino acids, carbohydrates, minerals, fatty acids and vitamins.
Oncogene	:	Gene implicated in the malignant tranformation of cells, and thus also implicated in cancer
Peptide	:	A class of compounds of low molecular weight which consist of two or more amino acids.
Peripheral blood cells	:	Those cells circulating in the bloodstream
Pharmacodynamics	:	The biochemical and physiological effects of drugs and the mechanisms of their actions.
Pharmacokinetics	:	The reaction of the body on a drug over a period of time, including processes of absorption, distribution, localization in tissues, biotransformation and excretion.
Phenotype	:	The expressed characteristics arising from the genotype
Plasmid	:	Loop of DNA capable of extrachromosomal replication in bacteria
Ploypropylene	:	A type of plastic which can withstand autoclaving
Polystyrene	:	A type of plastic which cannot withstand autoclaving
Posttranslational modification	:	Changes to a protein occurring after the genetic code has been 'translated' into an amino acid sequence (protein). Examples are the folding of a protein, the secondary deletion of amino acids, the formation of structural bridges or the addition of small molecules (sugar, phosphates etc) to a protein.
Primary cell culture	:	A culture of cells, tissues or organs taken directly from organisms and before the first subculture.

Table (*Contd...*)

Table (*Contd...*)

Term		*Description*
Proliferation	:	The division of cells producing further cells of the same type
Protein	:	Large complex organic compounds built of many amino acids. They are the principle constituents of cells.
Quiescence	:	The resting state of a cell (G_0/G_1)
Receptor	:	A specific molecule on the surface or within the cytoplasm of a cell that specifically recognizes and binds other molecules, *e.g.* antigens, hormones or neurotransmitter molecules. The specific binding elicits a specific response (reaction).
Retrovirus	:	A type of virus using RNA as its genetic material
Senescence	:	Cell or cells which have lost their growth potential
Serial dilution	:	A series of repetitive dilutions
Somatic cell therapy	:	The administration of cells grown or treated in culture to a patient to cure or alleviate the symptoms of a disease.
Sterilize	:	Remove microbial contaminants
Sub-culturing	:	To perpetuate a cell line by reseeding a small number of cells into fresh medium
Substratum	:	The surface upon which cells can rest or attach while growing
Suspension culture	:	Cells which grow without the need for adherence on to substratum
Syngeneic	:	Animals which have been inbred to produce a genetically similar/ identical population
Systemic toxicity	:	Damaging effects to the body as a whole.
Titre	:	Concentration of infective agents
Transfection	:	Introduction of foreing DNA into a cell
Transfection	:	The process of transferring or introducing a gene or nucleic acid to a recipient cell.
Transform	:	To transfect a bacterial cell, also to induce a cell to express abnormal characteristics (often similar to those found in tumor cells)
Transgenic organism	:	An organism living with an altered or additional gene as a result of genetic engineering.
Trypsinization	:	To use the enzyme trypsin to remove adherence proteins from cell surfaces.

SUBJECT INDEX

D

E

F

G

H

I

K

L

M

N

O

P

Q